PRACTICE
Workbook

Grade 3

Harcourt
SCHOOL PUBLISHERS

Visit *The Learning Site!*
www.harcourtschool.com

TEXAS HSP Math

2 3 4 5 6 7 8 9 10 073 16 15 14 13 12 11 10 09 08

Contents

UNIT 4: MULTIPLICATION CONCEPTS AND FACTS

UNIT 5: DIVISION CONCEPTS AND FACTS

UNIT 6: GEOMETRY

UNIT 7: MEASUREMENT

UNIT 8: FRACTIONS, MULTIPLICATION, AND PROBABILITY

© Harcourt

Ways to Use Numbers

Tell how each number is used. Write *count, measure, label,* or *position*.

1.

2.

3.

15 marbles

_____ _____ _____

4. Shanice placed 4th at the Texas state swim meet.

5. Robby has 75 hangers hanging in his closet.

_____

_____

6. Drew is 5 feet tall.

7. Audrey's locker number is #364.

_____

_____

8. The mall is located at 327 Commerce Boulevard.

9. Meryl's physical fitness scores put her in the 95th percentile.

_____

_____

Problem Solving and TAKS Prep

10. The state of Texas has the 2nd highest population in the United States. How is 2nd used in this fact?

11. Texas's highest point is Guadalupe Peak, which stands 8,749 feet tall. How is 8,749 used in this fact?

_____

_____

12. Which number is used to count?

A 65 third graders

B 2nd place winner

C 100-pound pumpkin

D Office Suite

13. Mr. Carlson lives at 418 Laurel Lane in Dallas, Texas. How is 418 used here?

F to count

G to label

H to measure

J to position

© Harcourt

Practice

Algebra: Patterns on a Hundred Chart

Use the hundred chart. Find the next number in the pattern.

1	2	3	4	5	6	7	8	9	10
11	12	13	14	15	16	17	18	19	20
21	22	23	24	25	26	27	28	29	30
31	32	33	34	35	36	37	38	39	40
41	42	43	44	45	46	47	48	49	50
51	52	53	54	55	56	57	58	59	60
61	62	63	64	65	66	67	68	69	70
71	72	73	74	75	76	77	78	79	80
81	82	83	84	85	86	87	88	89	90
91	92	93	94	95	96	97	98	99	100

1. 1, 3, 5, 7, ____

2. 6, 5, 4, 3, ____

3. 10, 15, 20, 25, ____

4. 15, 12, 9, 6, ____

5. 10, 20, 30, 40, ____

6. 65, 63, 61, 59, ____

Use the hundred chart. Tell whether each number is
odd or *even*.

7. 7 _____

8. 36 _____

9. 50 _____

10. 77 _____

11. 98 _____

12. 90 _____

13. 8 _____

14. 24 _____

15. 21 _____

16. 33 _____

17. 9 _____

18. 85 _____

Practice

Locate Points on a Number Line

Find the number that point *X* stands for on the number line.

1.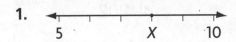

5   X   10

_____

2.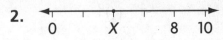

0   X   8   10

_____

3.

12   15   X   27   30

_____

4.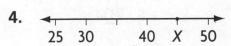

25   30   40   X   50

_____

Problem Solving and TAKS Prep

For 5–6, use the number line below.

70   S   90   X   110

5. Raul's trivia score is shown by point *S*. What is Raul's score?

_____

6. Raul answered two more questions correctly. His new score is labeled point *X*. What is Raul's new score?

_____

For 7–8, use the number line below.

3   6   R   S   T   U   21   24

7. Which point represents the number 18 on the number line?

A *R*

B *T*

C *S*

D *U*

8. Which number does point *R* represent?

F 9

G 12

H 15

J 25

© Harcourt

Practice

Place Value: 3 Digits

Write the value of the underlined digit.

1. 81<u>8</u>

2. 1<u>9</u>1

3. <u>8</u>17

4. <u>9</u>02

_____

_____

_____

_____

5. 25<u>3</u>

6. 7<u>0</u>4

7. 6<u>4</u>0

8. <u>3</u>97

_____

_____

_____

_____

Write each number in standard form.

9. 300 + 40 + 2

10. 500 + 60 + 1

11. 200 + 10 + 9

_____

_____

_____

12. seven hundred three

13. four hundred ninety-nine

_____

_____

Write each number in expanded form.

14. 921

15. 650

16. two hundred fifty

_____

_____

_____

Problem Solving and TAKS Prep

17. Female elk can weigh up to six hundred pounds. In standard form, how many digits that are not zeros does this weight contain?

18. Male mountain lions can weigh up to one hundred sixty pounds. In a place value chart of this weight, which digit would be positioned in the hundreds place?

_____

_____

19. Which shows six hundred five written in expanded form?

   A 605   **C** 600 + 5

   B 650   **D** 600 + 50

20. Which shows four hundred forty written in standard form?

   F 400   **H** 444

   G 440   **J** 400 + 40

Practice

Place Value: 4 Digits

Write each number in standard form.

1. $9,000 + 8$

_____

2. six thousand, one hundred twelve

_____

3. four thousand, two hundred two

_____

4. $2,000 + 700 + 30 + 4$

_____

Write each number in expanded form.

5. 3,724

_____

6. 5,209

_____

7. 6,009

_____

8. 9,638

_____

9. seven thousand four

_____

10. four hundred seventy-seven

_____

Write the value of the underlined digit.

11. 9,876

_____

12. 7,219

_____

13. 3,147

_____

14. 4,296

_____

Problem Solving and TAKS Prep

15. Write a 4-digit number that contains the digits 0, 1, 2, and 3. What is the value of the first digit in your number?

_____

16. Harry will eat 1,500 peanut butter and jelly sandwiches before he graduates from college. How would you write 1,500 in word form?

_____

17. Which number shows five thousand, three hundred two?

   A 532   C 5,302

   B 5,032   D 5,320

18. Which is the value of the underlined digit in 7,318?

   F 7   H 700

   G 70   J 7,000

Place Value: 5 and 6 Digits

Write the value of the underlined digit.

1. 3̲4,219

2. 7̲28,516

3. 15̲6,327

4. 4̲05,318

_____ _____ _____ _____

5. 211,0̲07

6. 80,23̲9

7. 44,9̲20

8. 3̲00,999

_____ _____ _____ _____

Write each number in standard form.

9. 70,000 + 8,000 + 300 + 5

10. forty-three thousand, eleven

_____ _____

11. 900,000 + 60,000 + 20 + 6

12. three hundred seventy-three thousand, eight hundred sixty-one

_____ _____

Problem Solving and TAKS Prep

13. Mauna Kea, an inactive volcano in Hawaii, stands 13,796 feet above sea level. What is the value of the digit 1 in 13,796?

14. During the one-week period it was held, 237,465 people attended the state fair. How would you write 237,465 in word form?

_____

_____ _____

15. Which number is the greatest?

   A 24,030
   B 24,300
   C 24,330
   D 24,000

16. Which is the value of the digit 9 in 987,654?

   F 9
   G 900
   H 9,000
   J 900,000

Practice

Problem Solving Workshop Strategy: Use Logical Reasoning

Problem Solving Strategy Practice

Use logical reasoning to solve.

1. Mario's locker number is between 80 and 99. The sum of the digits is 13. The tens digit is 3 more than the ones digit. What is Mario's locker number?

_____

2. In a spelling bee, Cal, Dawn, and Amy were the top three finishers. Cal finished in second place. Dawn did not finish first. Who finished first?

_____

3. Eight students tried out for the band or the chorus. Five students tried out for the band, the rest tried out for the chorus. How many students tried out for the chorus?

_____

4. Earl answered 2 more math questions correctly than Anna did. Anna answered 3 fewer questions correctly than Juanita did. Juanita answered 21 questions correctly. How many questions did Earl answer correctly?

_____

Mixed Strategy Practice

5. Doug has 170 stamps in his collection. His first book of stamps has 30 more stamps in it than his second book. How many stamps are in each book?

_____

6. Mr. Burns ran 14 miles last week. He only ran on Monday, Tuesday, and Wednesday. If Mr. Burns ran 3 miles on Tuesday and 5 miles on Wednesday, then how many miles did he run on Monday?

_____

7. **USE DATA** Josie is 2 spans plus 3 ells tall. How many feet tall is Josie?

_____

Unusual Measurements
1 fathom = 6 feet
2 spans = 3 feet
3 ells = 1 foot

Practice

Compare Numbers

Compare the numbers. Write <, >, or = for each ◯.

1. 78 ◯ 87

2. 100 ◯ 99

3. 529 ◯ 592

4. 84 ◯ 84

5. 964 ◯ 946

6. 624 ◯ 642

7. 297 ◯ 97

8. 173 ◯ 317

9. 321 ◯ 312

10. 94 ◯ 940

11. 724 ◯ 724

12. 239 ◯ 29

13. 870 ◯ 87

14. 638 ◯ 863

15. 574 ◯ 745

16. 746 ◯ 746

17. 404 ◯ 374

18. 393 ◯ 403

19. 632 ◯ 632

20. 206 ◯ 204

Problem Solving and TAKS Prep

21. **Fast Fact** The tallest building in Dallas is the Bank of America Plaza, which stands 921 feet tall. The tallest building in Albuquerque is Albuquerque Plaza, which stands 315 feet tall. Compare the heights of these two buildings.

_____

22. The 3rd grade has 384 students in it. The 4th grade has 348 students in it. Compare the numbers of students in these two grade levels.

_____

23. Which number is less than 952 but greater than 924?

A 925

B 952

C 955

D 1,000

24. Which number is greater than 786 but less than 791?

F 678

G 768

H 786

J 790

Practice

Order Numbers

Write the numbers in order from least to greatest.

1. 12, 92, 32

2. 37, 34, 39

3. 86, 88, 85

_____

_____

_____

4. 500, 300, 400

5. 139, 142, 127

6. 587, 583, 582

_____

_____

_____

Write the numbers in order from greatest to least.

7. 39, 27, 58

8. 82, 89, 91

9. 76, 74, 78

_____

_____

_____

10. 218, 312, 199

11. 652, 671, 649

12. 437, 439, 436

_____

_____

_____

13. 562, 526, 625

14. 987, 978, 998

15. 249, 429, 294

_____

_____

_____

Problem Solving and TAKS Prep

16. Fast Fact The Sabine River is 555 miles long. The Neches River is 416 miles long. The Trinity River is 508 miles long. Write the names of the rivers in order from least to greatest length.

_____

_____

17. Reasoning I am a number that is greater than 81 but less than 95. The sum of my digits is 15. What number am I?

18. Which number is the greatest?

   A 536   **C** 653

   B 635   **D** 563

19. Which number is greater than 498 but less than 507?

   F 497   **H** 507

   G 499   **J** 510

Compare and Order Greater Numbers

Write the numbers in order from least to greatest.

1. 587; 578; 5,087   **2.** 2,315; 2,135; 2,531   **3.** 3,721; 3,735; 3,719

_____   _____   _____

4. 4,001; 4,100; 420   **5.** 5,718; 3,718; 1,718   **6.** 8,239; 8,199; 8,098

_____   _____   _____

Write the numbers in order from greatest to least.

7. 913; 1,013; 1,031   **8.** 6,329; 6,239; 6,392   **9.** 7,428; 7,425; 7,429

_____   _____   _____

10. 5,230; 3,250; 2,350   **11.** 9,909; 999; 9,099   **12.** 5,768; 5,876; 5,687

_____   _____   _____

Problem Solving and TAKS Prep

USE DATA For 13–14, use the table below.

13. Which player has the greatest number of pass completions?

_____

14. Write the numbers of pass completions from the table in order from least to greatest.

_____

Football Hall of Fame	
Name	Pass Completions
Troy Aikman	2,742
Don Meredith	1,170
Sammy Baugh	1,754

15. Which number is the greatest?

    A 8,327

    B 8,273

    C 8,372

    D 8,237

16. Which number is less than 4,726 but greater than 3,998?

    F 3,997

    G 3,999

    H 4,726

    J 4,727

Practice

Round to the Nearest Ten

Round the number to the nearest ten.

1. 52   **2.** 47   **3.** 95   **4.** 107   **5.** 423

_____   _____   _____   _____   _____

6. 676   **7.** 209   **8.** 514   **9.** 673   **10.** 19

_____   _____   _____   _____   _____

11. 478   **12.** 313   **13.** 627   **14.** 789   **15.** 204

_____   _____   _____   _____   _____

Problem Solving and TAKS Prep

USE DATA For 16–17, use the table below.

16. To the nearest ten, what was the number of sea lions spotted on Friday?

_____

17. To the nearest ten, what was the number of sea lions spotted in all?

_____

Sea Lions Spotted Off the Pier	
Day	Number of Sea Lions Spotted
Friday	48
Saturday	53
Sunday	65

18. The number of stamps in Krissy's collection, rounded to the nearest ten, is 670. How many stamps could Krissy have?

A 679

B 676

C 669

D 664

19. On a number line, point X is closer to 350 than it is to 360. Which number could point X be?

F 354

G 356

H 361

J 365

Practice

Round to the Nearest Hundred

Round the number to the nearest hundred.

1. 349   **2.** 251   **3.** 765   **4.** 3,218   **5.** 6,552

_____   _____   _____   _____   _____

6. 4,848   **7.** 5,298   **8.** 6,342   **9.** 7,112   **10.** 412

_____   _____   _____   _____   _____

11. 901   **12.** 5,451   **13.** 2,982   **14.** 9,216   **15.** 1,543

_____   _____   _____   _____   _____

Problem Solving and Test Prep

USE DATA For 16–17, use the table below.

16. To the nearest hundred, how many feet tall is Texas' highest point?

_____

17. To the nearest hundred, how many miles long is Texas' border with Mexico?

_____

Texas Geography	
Feature	**Size**
Border with Mexico	1,001 miles
Highest Point	8,751 feet
Rio Grande	1,885 miles long

18. Which number does NOT round to 500, when rounded to the nearest hundred?

  A 450

  B 499

  C 533

  D 552

19. On a number line, point P is closer to 300 than to 200. Which number could point P stand for?

  F 219

  G 247

  H 273

  J 202

     Practice

Problem Solving Workshop Skill: Use a Number Line

Problem Solving Skill Practice

For 1–4, use the number line and the animal weight data.

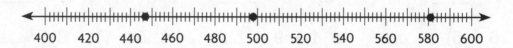

The zoo has a lion that weighs 447 pounds, a zebra that weighs 498 pounds, and a grizzly bear that weighs 581 pounds.

1. Is the weight of the zebra closer to 400, or 500, pounds?

2. To the nearest hundred, what is the weight of the lion?

3. To the nearest hundred, what is the weight of the grizzly bear?

4. To the nearest hundred, how much do the animals weigh in all?

Mixed Applications

5. Charlie collected 5 marbles on Monday, 3 marbles on Wednesday, and 2 marbles on Tuesday. How could you put the numbers of marbles Charlie collected in order based on the days that they were collected from earliest to latest?

6. Patrick brought 4 pencils to school on Monday, 3 pencils to school on Tuesday, and a bagged lunch to school on Wednesday. How many pencils did Patrick bring to school on Monday and Tuesday combined?

7. Luke put three numbers in order from least to greatest. The total amount of digits in the numbers he ordered is 4. Are any of the three numbers Luke put into order made up of more than 2 digits?

8. Lana won two spelling bees last year. She told her mother that in the number 1,020 the number in the hundreds place has a value of 0. Is what Lana told her mother correct?

Practice

Algebra: Addition Properties

Find each sum.

1. $4 + 7 =$ _____ **2.** $1 + (8 + 5) =$ _____ **3.** $(3 + 9) + 4 =$ _____

  $7 + 4 =$ _____ $(1 + 8) + 5 =$ _____ $3 + (9 + 4) =$ _____

4. $4 + (6 + 6) =$ _____ **5.** $1 + 9 =$ _____ **6.** $5 + (3 + 3) =$ _____

  $(4 + 6) + 6 =$ _____ $9 + 1 =$ _____ $(5 + 3) + 3 =$ _____

Find each sum in two different ways. Use parentheses to show which numbers you added first.

7. $7 + 3 + 5 =$ _____   **8.** $9 + 4 + 2 =$ _____

9. $62 + 18 + 5 =$ _____   **10.** $25 + 4 + 6 =$ _____

11. $1 + 42 + 9 =$ _____ **12.** $0 + 16 + 16 =$ _____ **13.** $9 + 7 + 9 =$ _____

14. $14 + 6 + 3 =$ _____ **15.** $50 + 6 + 30 =$ _____ **16.** $21 + 42 + 1 =$ _____

Problem Solving and TAKS Prep

17. On a nature walk, Sarah sees 3 squirrels, 5 chipmunks, and 8 birds. How many animals does Sarah see in all?

_____

18. On Monday Ramon saw 4 squirrels and 8 birds, in the park. On Tuesday he saw 8 squirrels and 4 birds, in the park. On Monday and Tuesday, how many animals did Ramon see in all?

_____

19. Which is the sum?
  $3 + 10 =$ _____

  A 0   **C** 13

  B 3   **D** 30

20. Which property is shown in the number sentence below?
  $8 + (9 + 4) = (8 + 9) + 4$

  F zero   **H** identity

  G commutative **J** associative

Algebra: Missing Addends

Find the missing addend. You may want to use counters.

1. $3 + \boxed{} = 10$ **2.** $\boxed{} + 9 = 14$ **3.** $\boxed{} + 6 = 11$ **4.** $\boxed{} + 2 = 5$

5. $\boxed{} + 7 = 13$ **6.** $2 + \boxed{} = 4$ **7.** $\boxed{} + 9 = 12$ **8.** $9 + \boxed{} = 17$

9. $6 + \boxed{} = 12$ **10.** $\boxed{} + 1 = 10$ **11.** $3 + \boxed{} = 8$ **12.** $\boxed{} + 4 = 4$

Find the missing number. You may want to use counters.

13. $9 + 9 = \underline{}$ **14.** $3 + \boxed{} = 12$ **15.** $5 + 5 = \underline{}$ **16.** $7 + 0 = \underline{}$

17. $6 + 8 = \underline{}$ **18.** $2 + \boxed{} = 10$ **19.** $\boxed{} + 5 = 12$ **20.** $\boxed{} + 0 = 3$

21. $8 + \boxed{} = 12$ **22.** $4 + 7 = \underline{}$ **23.** $6 + \boxed{} = 11$ **24.** $2 + 7 = \underline{}$

Problem Solving and TAKS Prep

25. Fast Fact A squirrel can run 12 miles per hour. A house mouse can run 8 miles per hour. How many miles per hour faster can a squirrel run than a house mouse can run?

_____

26. Sophia went to an amusement park. She went on 18 rides in all. Seven of the rides Sophia went on were roller coasters. How many rides that Sophia went on were not roller coasters?

_____

27. Which is the missing number?
   $2 + 7 = \underline{}$
   A 5
   B 6
   C 8
   D 9

28. Which is the missing addend for
   $11 + \underline{} = 15?$
   F 3
   G 4
   H 5
   J 6

Estimate Sums

Use rounding to estimate each sum.

1.	64 + 29	2.	45 + 21	3.	14 + 37	4.	423 + 17	5.	661 + 32

6.	271 + 349	7.	535 + 183	8.	721 + 248	9.	183 + 134	10.	387 + 97

Use compatible numbers to estimate each sum.

11.	48 + 34	12.	24 + 27	13.	17 + 64	14.	123 + 76	15.	572 + 25

16.	624 + 173	17.	804 + 136	18.	217 + 254	19.	345 + 453	20.	638 + 243

Problem Solving and TAKS Prep

USE DATA For 21–22, use the table below.

21. About how many different species of parrots and raptors are there?

_____

22. Which is greater, the estimated sum of pigeon and raptor species or the estimated sum of parrot and penguin species?

_____

Number of Bird Species	
Type of Bird	Number of Different Species
parrots	353
raptors	307
penguins	17
pigeons	309

23. A family drove 325 miles one day and 189 miles the next day. About how many miles did the family drive in all?

   A 50   **C** 500

   B 600   **D** 400

24. While walking around a lake, Toby counted birds. He counted 23 herons and 45 ducks. About how many herons and ducks did Toby count in all?

   F 100   **H** 50

   G 70   **J** 170

© Harcourt

Practice

Add 2-Digit Numbers

Estimate. Then find each sum using place value or mental math.

1.	2.	3.	4.	5.
19 + 64	33 28 + 14	63 + 45	34 + 76	65 48 + 16

6.	7.	8.	9.	10.
75 + 47	31 + 86	47 + 25	24 32 + 18	47 24 + 52

11. $56 + 41 =$ _____   **12.** $83 + 15 =$ _____   **13.** $25 + 67 + 31 =$ _____

14. $29 + 67 =$ _____   **15.** $37 + 21 =$ _____   **16.** $49 + 34 + 61 =$ _____

Problem Solving and TAKS Prep

17. Kara bought 13 green apples and some red apples. She bought a total of 40 apples. How many red apples did Kara buy?

18. Manuel and his brother picked apples. Manuel picked 62 apples. His brother picked 39 apples. How many apples did Manuel and his brother pick in all?

_____

_____

19. Which is the sum?

$71 + 23 + 18 =$ _____

A 89   **C** 102

B 94   **D** 112

20. Which is the sum?

$65 + 28 =$ _____

F 83   **H** 93

G 92   **J** 98

Practice

Model 3-Digit Addition

Use base-ten blocks to find each sum.

1. $128 + 356 = $ _____   **2.** $147 + 266 = $ _____   **3.** $594 + 245 = $ _____

4. $649 + 248 = $ _____   **5.** $392 + 455 = $ _____   **6.** $288 + 477 = $ _____

7. $388 + 256 = $ _____   **8.** $133 + 267 = $ _____   **9.** $818 + 103 = $ _____

Find each sum.

10. 821
  $+143$

11. 765
  $+154$

12. 217
  $+265$

13. 291
  $+645$

14. 608
  $+154$

15. 309
  $+512$

16. 485
  $+180$

17. 789
  $+101$

18. 236
  $+319$

19. 167
  $+418$

20. 189
  $+178$

21. 248
  $+318$

22. 378
  $+147$

23. 320
  $+575$

24. 256
  $+127$

25. 444
  $+328$

26. 701
  $+199$

27. 225
  $+387$

28. 821
  $+143$

29. 765
  $+154$

30. 635
  $+364$

31. 528
  $+122$

32. 137
  $+303$

33. 412
  $+101$

34. 862
  $+112$

Practice

Add 3-Digit Numbers

Estimate. Then find each sum.

1. 205
    +582

2. 725
    +237

3. 317
    +445

4. 377
    +429

5. 199
    +534

6. 627
    +312

7. 336
    +248

8. 743
    +185

9. 812
    +309

10. 476
    +358

11. 503
    258
    +507

12. 883
    399
    +174

13. 612
    483
    +744

14. 975
    194
    +585

15. 109
    237
    +176

16. $832 + 415 =$ _____

17. $358 + 329 =$ _____

18. $212 + 688 =$ _____

Problem Solving and TAKS Prep

19. Margie drove 665 miles from her home in Lubbock to her aunt's home in Brownsville for a vacation. She then drove the same distance to return home. How many miles did Margie drive in all?

20. Shawn has climbed 697 steps of the Eiffel Tower. He has 968 steps left to climb to reach the top. How many steps are on the Eiffel Tower?

_____

_____

21. Which is the sum of 467 and 384?

    A 741
    B 751
    C 841
    D 851

22. Which is the sum of 593 and 252?

    F 745
    G 755
    H 845
    J 855

Practice

Problem Solving Workshop Strategy: Guess and Check

Problem Solving Strategy Practice

1. There were 300 people at the football game. There were 60 more students than adults at the game. How many students were at the football game?

2. The gym coach ordered 56 total basketballs and soccer balls for next year. There were 10 fewer basketballs ordered than soccer balls ordered. How many of each type of ball were ordered?

Mixed Strategy Practice

USE DATA For 3–4, use the table.

3. Sami and Juan had the same number of baseball cards. Then Sami received some baseball cards as a present. How many cards did Sami receive as a present?

Baseball Cards Collected	
Name	Number of Cards
Sami	250
Pete	150
Juan	200

4. Pete has 50 baseball cards of players that are pitchers. He has 25 baseball cards of players that are catchers. The rest of his baseball cards are of players that are outfielders. How many of Pete's cards are of players that are outfielders?

5. Tom spent $35 on a new helmet and kneepads. He spent $15 for a new football. At the end of the day he had $5 left. How much money did Tom have to start?

6. Sarah, Jose, and Mike are sitting in a row. If you face them, Mike is not sitting on the left. Sarah is sitting to the right of Jose. Who is sitting in the middle?

Choose a Method

Find the sum. Tell which method you used.

1.	2.	3.	4.	5.
518 +220	422 +315	239 +521	679 +295	954 +756

6.	7.	8.	9.	10.
726 +384	231 +765	923 +855	523 +365	402 +509

11.	12.	13.	14.	15.
229 325 +558	904 675 +243	163 +741	239 +761	118 583 +236

16. 632 + 345 = _____ **17.** 192 + 153 = _____ **18.** 244 + 328 = _____

Problem Solving and TAKS Prep

19. A farmer planted 510 corn plants and 481 potato plants in his fields. How many corn and potato plants did the farmer plant in all?

_____

20. A farmer plants 615 tomato plants and 488 cucumber plants. How many plants does the farmer plant in all? Is mental math the best method to solve this problem? Explain.

_____

_____

21. Caroline's family plants 275 acres of corn and 386 acres of lettuce. How many acres total, does Caroline's family plant?

A 761

B 661

C 561

D 551

22. On Saturday, Jim, his father, and his two brothers harvested 304 acres. On Sunday they harvested 255 acres. How many acres did they harvest on both days combined? Explain which method you used.

_____

_____

© Harcourt

Algebra: Fact Families

Complete.

1. $6 - 4 = 2$, so $2 + \boxed{} = 6$

2. $3 + 8 = 11$, so $11 - \boxed{} = 3$

3. $12 - 9 = 3$, so $9 + \boxed{} = 12$

4. $7 + 6 = 13$, so $6 + \boxed{} = 13$

5. $8 + 8 = 16$, so $16 - \boxed{} = 8$

6. $17 - 9 = 8$, so $8 + \boxed{} = 17$

Write the fact family for each set of numbers.

7. 7, 8, 15

8. 5, 3, 8

9. 9, 9, 18

10. 6, 7, 13

11. 3, 7, 10

12. 7, 7, 14

Problem Solving and TAKS Prep

13. Kara is making muffins. She has 12 eggs. She uses 2 eggs to make the muffins. How many eggs does Kara have left?

14. **Reasoning** How can you use $7 + 4 = 11$ to find the missing number in $11 - \boxed{} = 4$?

15. Which number sentence is in the same fact family as $6 + 5 = 11$?

    A $6 - 5 = 1$ C $11 + 5 = 16$

    B $11 - 5 = 6$ D $7 + 4 = 11$

16. Which set of numbers can make a fact family?

    F 3, 4, 7   H 2, 3, 8

    G 5, 7, 11   J 4, 6, 9

Practice

Estimate Differences

Use rounding or compatible numbers to estimate each difference.

1. 74	2. 52	3. 47	4. 65
-38	-26	-13	-32

5. 371	6. 974	7. 721	8. 283
-159	-126	-358	-154

9. 978	10. 357	11. 787	12. 549
-447	-197	-268	-324

ALGEBRA. Estimate to compare. Write $<$, $>$, or $=$ for each $\bigcirc$.

13. $55 - 29 \bigcirc 50$ 14. $593 - 129 \bigcirc 300$ 15. $805 - 250 \bigcirc 500$

Problem Solving and TAKS Prep

USE DATA For 16–17 use the table below.

16. About how much more is the weight of the white sturgeon than the combined weight of the gar and the blue catfish?

_____

17. About how much more did the white sturgeon weigh than the Nile perch weighed?

_____

Largest Freshwater Fish Caught	
Type of Fish	Weight in Pounds
Gar	279
Nile Perch	213
Blue Catfish	111
White Sturgeon	468

18. Tammy estimated $923 - 452$. She rounded each number to the nearest hundred and then subtracted. Which was Tammy's estimate?

   A 300   C 500

   B 400   D 600

19. Which is the estimated difference? 659
    -382

   F 300   H 500

   G 400   J 600

Practice

Subtract 2-Digit Numbers

Estimate. Then find each difference.

1. $\begin{array}{r} 79 \\ -53 \\ \hline \end{array}$ 2. $\begin{array}{r} 35 \\ -14 \\ \hline \end{array}$ 3. $\begin{array}{r} 63 \\ -45 \\ \hline \end{array}$ 4. $\begin{array}{r} 76 \\ -58 \\ \hline \end{array}$ 5. $\begin{array}{r} 55 \\ -16 \\ \hline \end{array}$

6. $\begin{array}{r} 82 \\ -47 \\ \hline \end{array}$ 7. $\begin{array}{r} 68 \\ -31 \\ \hline \end{array}$ 8. $\begin{array}{r} 47 \\ -25 \\ \hline \end{array}$ 9. $\begin{array}{r} 97 \\ -19 \\ \hline \end{array}$ 10. $\begin{array}{r} 63 \\ -17 \\ \hline \end{array}$

Find each difference. Use addition to check.

11. $56 - 41 =$ _____

12. $83 - 35 =$ _____

13. $67 - 31 =$ _____

14. $36 - 19 =$ _____

15. $66 - 15 =$ _____

16. $91 - 22 =$ _____

Problem Solving and TAKS Prep

17. The brown bear has an average height of 48 inches. The American black bear has an average height of 33 inches. What is the difference between these two bear's average heights?

_____

18. An adult polar bear has a height of 63 inches. A polar bear cub has a height of 39 inches. What is the difference of these heights?

_____

19. Which is the difference?

$72 - 48 =$ _____

   A 24   **C** 34

   B 26   **D** 36

20. At a fair, a drink stand sold 45 glasses of lemonade and 29 glasses of tea. How many more glasses of lemonade than glasses of tea were sold?

   F 26   **H** 16

   G 24   **J** 14

Practice

Model 3-Digit Subtraction

Use base-ten blocks to find each difference.

1. $494 - 271 = $ _____

2. $324 - 147 = $ _____

3. $549 - 255 = $ _____

4. $311 - 205 = $ _____

5. $757 - 483 = $ _____

6. $623 - 197 = $ _____

7. $388 - 265 = $ _____

8. $267 - 183 = $ _____

9. $706 - 258 = $ _____

Find each difference.

10.
$$\begin{array}{r} 765 \\ -154 \\ \hline \end{array}$$

11.
$$\begin{array}{r} 821 \\ -143 \\ \hline \end{array}$$

12.
$$\begin{array}{r} 665 \\ -327 \\ \hline \end{array}$$

13.
$$\begin{array}{r} 821 \\ -581 \\ \hline \end{array}$$

14.
$$\begin{array}{r} 387 \\ -198 \\ \hline \end{array}$$

15.
$$\begin{array}{r} 309 \\ -212 \\ \hline \end{array}$$

16.
$$\begin{array}{r} 485 \\ -276 \\ \hline \end{array}$$

17.
$$\begin{array}{r} 784 \\ -359 \\ \hline \end{array}$$

18.
$$\begin{array}{r} 319 \\ -236 \\ \hline \end{array}$$

19.
$$\begin{array}{r} 418 \\ -276 \\ \hline \end{array}$$

20.
$$\begin{array}{r} 189 \\ -178 \\ \hline \end{array}$$

21.
$$\begin{array}{r} 548 \\ -318 \\ \hline \end{array}$$

22.
$$\begin{array}{r} 707 \\ -629 \\ \hline \end{array}$$

23.
$$\begin{array}{r} 845 \\ -563 \\ \hline \end{array}$$

24.
$$\begin{array}{r} 956 \\ -127 \\ \hline \end{array}$$

25.
$$\begin{array}{r} 752 \\ -382 \\ \hline \end{array}$$

26.
$$\begin{array}{r} 607 \\ -199 \\ \hline \end{array}$$

27.
$$\begin{array}{r} 387 \\ -225 \\ \hline \end{array}$$

28.
$$\begin{array}{r} 900 \\ -459 \\ \hline \end{array}$$

29.
$$\begin{array}{r} 765 \\ -150 \\ \hline \end{array}$$

30.
$$\begin{array}{r} 777 \\ -444 \\ \hline \end{array}$$

31.
$$\begin{array}{r} 228 \\ -116 \\ \hline \end{array}$$

32.
$$\begin{array}{r} 939 \\ -540 \\ \hline \end{array}$$

33.
$$\begin{array}{r} 442 \\ -378 \\ \hline \end{array}$$

34.
$$\begin{array}{r} 808 \\ -102 \\ \hline \end{array}$$

Practice

Subtract 3-Digit Numbers

Estimate. Then find the difference.

1. 593
   -282

2. 377
   -188

3. 732
   -489

4. 654
   -386

5. 534
   -175

6. 657
   -132

7. 673
   -583

8. 820
   -649

9. 812
   -309

10. 976
    -267

11. 578
    -126

12. 738
    -644

13. 472
    -281

14. 872
    -125

15. 477
    -298

Problem Solving and TAKS Prep

16. **Fast Fact** The Millennium Force roller coaster is 310 feet tall. The Goliath roller coaster is 235 feet tall. How many feet taller is the Millennium Force roller coaster than the Goliath roller coaster?

_____

17. The roller coaster Kingda Ka's steepest drop is 418 feet. The Goliath roller coaster's steepest drop is 163 feet less steep than Kingda Ka's drop. How steep is Goliath's steepest drop?

_____

18. Which is the difference between 945 and 194?

   A 651
   B 741
   C 751
   D 851

19. Which is the difference?

   852
   -374

   F 522
   G 488
   H 482
   J 478

Practice

Subtract Across Zeros

Estimate. Then find the difference.

1. $\begin{array}{r} 508 \\ -175 \\ \hline \end{array}$	2. $\begin{array}{r} 400 \\ -329 \\ \hline \end{array}$	3. $\begin{array}{r} 980 \\ -246 \\ \hline \end{array}$	4. $\begin{array}{r} 806 \\ -493 \\ \hline \end{array}$	5. $\begin{array}{r} 700 \\ -123 \\ \hline \end{array}$
6. $\begin{array}{r} 608 \\ -169 \\ \hline \end{array}$	7. $\begin{array}{r} 701 \\ -213 \\ \hline \end{array}$	8. $\begin{array}{r} 408 \\ -184 \\ \hline \end{array}$	9. $\begin{array}{r} 930 \\ -429 \\ \hline \end{array}$	10. $\begin{array}{r} 500 \\ -379 \\ \hline \end{array}$

Find each difference. Use addition to check.

11. $902 - 426 =$ _____

12. $800 - 424 =$ _____

13. $600 - 431 =$ _____

14. $500 - 265 =$ _____

15. $408 - 225 =$ _____

16. $830 - 315 =$ _____

Problem Solving and TAKS Prep

17. Juan plays arcade games and wins some tickets. He needs 400 tickets for a beach ball. He already has 252 tickets. How many more tickets does Juan need?

_____

18. Hannah plays arcade games and wins 243 tickets. She needs 700 tickets for a sweatshirt. How many more tickets does Hannah need?

_____

19. Which is the difference?

$600 - 328 =$ _____

A 272
B 282
C 372
D 382

20. Which is the difference?

$\begin{array}{r} 806 \\ -238 \\ \hline \end{array}$

F 478
G 568
H 578
J 668

Practice

Choose a Method

Find the difference. Tell which method you used.

1.	518 −315	2.	732 −315	3.	925 −521	4.	659 −292	5.	945 −467

6.	922 −414	7.	675 −198	8.	800 −432	9.	635 −227	10.	509 −288

11.	909 −558	12.	954 −843	13.	632 −212	14.	569 −347	15.	418 −236

16. $755 - 172 =$ _____ **17.** $218 - 125 =$ _____ **18.** $784 - 318 =$ _____

Problem Solving and TAKS Prep

19. A polar bear at the zoo weighs 792 pounds. A giant panda at the zoo weighs 273 pounds. How many more pounds does the polar bear weigh than the giant panda weighs? Tell which method you used to solve.

_____

_____

20. A Cheetah can reach speeds of up to 66 miles per hour. A three-toed sloth can reach speeds of up to 8 miles per hour. What is the difference of these two speeds? Tell which method you used to solve.

_____

_____

21. An adult zebra weighs 725 pounds. An adult Siberian tiger weighs 562 pounds. How many more pounds does the zebra weigh than the Siberian tiger weighs?

A 163 pounds **C** 243 pounds

B 173 pounds **D** 263 pounds

22. A male cheetah weighs 142 pounds. A male panther weighs 121 pounds. What is the difference in weight between the cheetah and the panther?

F 121 pounds **H** 21 pounds

G 112 pounds **J** 12 pounds

Problem Solving Workshop Skill: Choose the Operation

Problem Solving Skill Practice

Tell which operation you would use. Then solve the problem.

1. Julia read 128 pages in a book. She needs to read 175 more pages to finish the book. How many pages total, are in the book?

   _____

   _____

2. The library has 325 books about animals. Of these, 158 are checked out. How many books about animals are still in the library?

   _____

   _____

3. Kara plans to put together a puzzle. The puzzle contains 225 pieces. She has put together 137 pieces. How many more pieces does Kara need to put together to complete the puzzle?

   _____

   _____

4. Jeremy had 529 coins in his collection. He collected 217 more coins. How many coins are now in Jeremy's collection?

   _____

   _____

Mixed Applications

USE DATA For 5–6, use the table below.

5. How many glasses of lemonade were sold in all, from Monday to Friday? Will you need to use an estimate or an exact answer?

   _____

6. On Saturday, the lemonade stand sold 15 glasses of lemonade. How many more glasses were sold on Friday than were sold on Wednesday?

   _____

Glasses of Lemonade Sold	
Day	**Number of Glasses Sold**
Monday	8
Tuesday	11
Wednesday	10
Thursday	7
Friday	15

7. The library contains 217 magazines, and 60 videos, that students can check out. Students have 109 magazines checked out. How many magazines are now available at the library?

   _____

Practice

Count Bills and Coins

Write the amount.

1.

2.

3.

4.

_____ _____ _____ _____

Find two equivalent sets for each. List the coins and bills.

5. $4.45

6. $1.58

7. 85¢

_____ _____ _____

_____ _____ _____

8. $3.25

9. 50¢

10. $6.50

_____ _____ _____

_____ _____ _____

Problem Solving and TAKS Prep

11. **Reasoning** Ana wants to buy a book for $3.95. List the fewest types and amounts of bills and coins Ana can use.

12. Wil has three $1 bills, 2 quarters, 3 dimes, a nickel, and 4 pennies. How much money does Wil have in all?

_____ _____

13. Mike has two $1 bills, 3 quarters, and 2 nickels. Shelley has 9 quarters, 3 dimes, and 11 pennies. How much more money does Mike have than Shelley has?

14. Mitch wants to buy a fruit salad. It costs $1.30. Which shows the fewest number of bills and coins Mitch can use?

| A 18 cents | C 19 cents | F 3 | H 5 |
| B 21 cents | D 20 cents | G 4 | J 6 |

Compare Money Amounts

Use <, >, or = to compare the amounts of money.

1.

2.

_____   _____

Which amount is greater?

3. $7.95 or
$7.89

4. $2.10 or
9 quarters

5. 87¢ or
18 nickels

6. 2 dimes,
6 pennies or
1 quarter

_____   _____

7. 2 dimes,
2 nickels or
1 quarter

8. $2.25 or
5 half dollars

9. 35 pennies or
2 dimes,
2 nickels, and
2 pennies

10. $1.12 or
11 dimes

_____   _____

Problem Solving and TAKS Prep

11. Aidan has 7 quarters, 3 dimes,
3 nickels, and 4 pennies. Maria
has 1 one-dollar bill, 1 half-dollar,
1 quarter, 2 dimes, and 3 nickels.
Who has more money, Aidan or
Maria? Explain.

12. Matt has 5 quarters, 6 dimes, and
4 nickels. Hal has $2.51. Who has
more money, Matt or Hal? Explain.

_____   _____

13. Becky has only dimes. She has
more than 60¢. Which amount
could Becky have?

    A 75¢   **C** 81¢

    B 50¢   **D** 70¢

14. Danny has only quarters and dimes.
He has at least 1 quarter and
1 dime. He has more than 25¢.
Which amount could Danny have?

    F 45¢   **H** 64¢

    G 30¢   **J** 40¢

Practice

Problem Solving Workshop Strategy: Make a Table

Problem Solving Strategy Practice

Elena has the bills and coins shown. She wants to buy a card for $2.95.

Fill in the table to show equivalent sets of $2.95.

$1 Bills	Quarters	Dimes	Nickels	Pennies	Total Value
1. _____	3	2	0	0	$2.95
2	2	4	**2.** _____	0	$2.95
2	**3.** _____	4	0	5	$2.95
1	7	2	0	**4.** _____	$2.95
1	6	4	**5.** _____	0	$2.95
1	**6.** _____	4	0	5	$2.95
1	7	1	1	5	$2.95

Mixed Strategy Practice

7. **USE DATA** What are the two most popular types of books? How do you know?

_____

_____

Favorite Type of Book	
Type	**Votes**
Sports	卌 ‖
Mystery	‖‖
Fantasy	卌 ‖
Science Fiction	‖‖‖

Practice

Model Making Change

Find the amount of change. Use play money and counting on to help.

1. Ali buys a collar for her dog for $4.69. She pays with a $10 bill.

_____

2. Roger buys a banana for $0.49. He pays with a $1 bill.

_____

Find the amount of change. List the coins and bills.

3. Pay with a $10 bill.

_____
_____
_____
_____
_____

$6.37

4. Pay with a $5 bill.

_____
_____
_____
_____

$2.79

Problem Solving and TAKS Prep

5. Isaac buys sunglasses for $3.99. He pays with one $10 bill. How much change does he receive? List the coins and bills.

_____
_____

6. Zoe wants to buy a clown wig that costs $5.99 and face paint that costs $1.15. She has $6.10. Is this enough money? If not, how much more does Zoe need?

_____

7. Sally buys a book that costs $3.54. She has $5.00. How much change does Sally receive?

   A $2.64   **C** $1.46

   B $1.54   **D** $8.54

8. Lori wants to buy a CD that costs $10.39. She has $7.50. How much more money does Lori need?

   F $2.89   **H** $17.89

   G $3.11   **J** $3.89

© Harcourt

Understand Time

Write the time. Then write two ways you can read the time.

1.

_____
_____

2.

_____
_____

3.

_____
_____

4.

_____
_____

For 5–8, write the letter of the clock that shows the time.

a.

b.

c.

d.

5. 25 minutes after 8 _____

6. 11:40 _____

7. 15 minutes before 3 _____

8. 2:45 _____

Problem Solving and TAKS Prep

9. Tim told Mark to meet him at exactly a quarter to ten. Mark arrived at 10:15. Did Mark meet Tim on time?

_____

10. At what times will Mary's digital clock display a one and three zeros for the time?

_____

11. Burt got up at quarter to seven. Which is one way to write this time?

   A 7:15   C 6:45

   B 7:45   D 6:15

12. Elena ate dinner at twenty minutes before six. Which is one way to write this time?

   F 5:40   H 6:40

   G 6:20   J 5:20

© Harcourt

Time to the Minute

Write the time. Then write one way you can read the time.

1.

2.

3.

4.

_____ _____ _____ _____

_____ _____ _____ _____

Problem Solving and TAKS Prep

5. Show the times: quarter past twelve, and forty-five minutes before one.

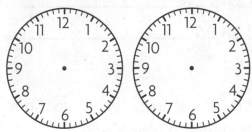

6. It's 29 minutes after 6. Show this time on the clock below.

7. Luis got up at twelve minutes before eight. What is one way to write this time?

8. What time is shown on this clock?

A 8:12   **C** 8:48

B 6:48   **D** 7:48

F 5:47   **H** 4:47

G 4:45   **J** 5:50

Minutes and Seconds

Write the time. Then write how you would read the time.

1.

2.

3.

4.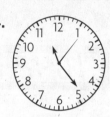

_____   _____   _____   _____

_____   _____   _____   _____

For 5–8, write the letter of the clock that shows the time.

a.

b.

c.

d.

5. 1:11:16 _____   6. 12:46:21 _____

7. 9:13:49 _____   8. 3:28:05 _____

Problem Solving and TAKS Prep

9. What is another way to say 60 seconds after 2:10?

_____

10. It is 29 seconds after 8:05. Show the time on May's watch.

11. Which time is 60 seconds before 3:00?

   A 2:59:00   C 3:01:00
   B 2:59:60   D 3:00:60

12. Which answer choice shows 6 seconds after 4:43?

   F 4:06:43   H 4:60:43
   G 4:43:60   J 4:43:06

Practice

A.M. or P.M.

Write the time for each activity. Use A.M. or P.M.

1. play basketball   **2.** eat lunch

3. go to the library   **4.** eat dinner

_____   _____   _____   _____

Write the time using numbers. Use A.M. or P.M.

5. eight twenty in the morning

6. five minutes after three in the afternoon

_____   _____

7. fifteen minutes before eleven at night   **8.** six forty-five in the morning

_____   _____

Problem Solving and TAKS Prep

9. Martha plays soccer every Saturday morning at 10 o'clock. Write this time using A.M. or P.M.

10. Debra plays soccer on Sunday mornings at twenty minutes to twelve. Write this time using A.M. or P.M.

_____   _____

11. At which times shown are most third graders awake?

    A 7:00 P.M.   **C** 3:00 A.M.

    B midnight   **D** 11:00 P.M.

12. At which times shown are most third graders asleep?

    F 7:00 P.M.   **H** 3:00 P.M.

    G midnight   **J** 11:00 A.M.

Problem Solving Workshop Skill:
Too Much/Too Little Information

Problem Solving Skill Practice

Tell whether there is too much, or too little, information. Solve if there is enough information.

1. Rochelle went to dance class at 8:00 A.M. What time did Rochelle's dance class end?

_____

_____

2. Jack has 3 bills, 2 quarters, and 1 dime in his pocket. He earned the money doing chores. Does Jack have enough money to buy a juice box that costs 55¢?

_____

_____

3. Caroline is in third grade. She arrives at the bus stop at 7:30 A.M. How long does Caroline have to wait for the bus?

_____

Mixed Applications

4. Teddy's soccer game starts at 4:15 P.M. He is the goalie. He makes 8 saves. The other goalie on his team makes 5 saves. How many saves were made in all? Choose the operation needed to solve.

_____

5. Stacy has 35 markers. She forgets to put the caps on 8 markers and they dry up. How many usable markers does Stacy have left? Choose the operation needed to solve.

_____

6. **USE DATA** Aidan goes swimming and then hiking for his morning activities. How much time does he spend on both activities?

_____

Camp Pine Activity Schedule		
Activity	Time	Length
Swimming	10:00 A.M.	30 minutes
Hiking	11:00 A.M	45 minutes

© Harcourt

Practice

Collect Data

For 1–4, use the Musical Instruments Students Play list.

1. Make a tally table to organize the data.

Musical Instruments Students Play	
Instrument	**Tally**

Musical Instruments Students Play

Jen	piano
Lisa	violin
Tarik	clarinet
Randy	piano
Jude	violin
Leah	guitar
Audry	clarinet
Sue	piano
Marty	violin
Debra	violin

2. How many students play the clarinet? _____

3. How many more students play the violin than play the guitar? _____

4. How many students in all, play musical instruments? _____

For 5–6, use the table below.

5. How many students play baseball?

_____

6. How many more students are involved in scouting than are involved in gymnastics?

_____

After-School Activities	
Activity	**Number of Students**
Baseball	7
Gymnastics	3
Scouting	8

Problem Solving and TAKS Prep

7. Seven students voted for vanilla ice cream and 4 for strawberry ice cream. How many students in all, voted for vanilla or strawberry ice cream?

_____

8. Four more students voted for soccer than voted for skating. If 12 students voted for soccer, how many students voted for skating?

_____

9. Which of the following numbers represents 卌 卌 卌 I ?

A 4

B 13

C 15

D 16

10. Which of the following is in order from least to greatest?

F 4, 卌, 6

G 6, 卌, 4

H 4, 5, 卌

J 卌, 5, 6

Practice

Read a Pictograph

For 1–3, use the pictograph below.

1. How many birds were sold on Thursday?

   _____

2. How many birds were sold from Thursday through Sunday?

   _____

3. On which two days combined were as many birds sold as on Saturday?

   _____

Number of Birds Sold

Thursday	🐦 🐦 🐦 🐦
Friday	🐦 🐦
Saturday	🐦 🐦 🐦 🐦
Sunday	🐦 🐦 🐦

Key: Each 🐦 = 3 birds.

Problem Solving and TAKS Prep

4. Sal visited 5 national parks last summer and 3 national parks this summer. How many national parks has Sal visited in the last two summers combined?

   _____

5. **Reasoning** A pictograph shows 🌳🌳🌳 to represent 12 parks. How many parks does 🌳 represent?

6. A pictograph uses the key 🐱 = 5 cats. Which picture represents 15 cats?

   A 🐱 🐱

   B 🐱 🐱 🐱

   C 🐱 🐱 🐱 🐱 🐱

   D 🐱 🐱 🐱 🐱 🐱 🐱

7. A pictograph uses the key ▮ = 10 gallons of water. How many gallons of water does ▮ ▮ ▮ stand for?

   F 2

   G 10

   H 25

   J 50

Practice

© Harcourt

Problem Solving Workshop Strategy:
Make a Graph

Problem Solving Strategy Practice

Make a pictograph to solve.

1. A group of students voted for their favorite farm animal. The results are shown below. Make a pictograph for the data. Let each picture stand for 4 votes.

Cow	8 votes
Chicken	12 votes
Horse	10 votes
Sheep	4 votes

Favorite Farm Animals	

Key: Each_____ = 4 votes.

2. If the key is changed so that each picture stands for 2 votes, then how many pictures should be used to represent the number of students who voted for horse?

_____

Mixed Strategy Practice

3. Chuck ate 2 pieces of watermelon before boarding the airplane. On the airplane, Chuck ate 3 bananas and a piece of steak. Then Chuck watched a movie. How many pieces of fruit did Chuck eat in all?

_____

4. Betty has a total of 19 dolls. She received one new doll from her aunt and 4 new dolls from other relatives. How many dolls did Betty have before she received the new dolls?

_____

© Harcourt

Practice

Read a Bar Graph

For 1–2, use the Time Spent Waiting in Line graph below.

1. How long was the wait to ride the bumper cars?

   _____

2. How much longer was the wait to ride the roller coaster than the wait to ride the Ferris wheel?

   _____

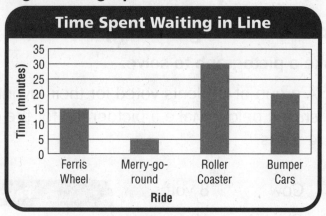

Problem Solving and TAKS Prep

For 3–4, use the bar graph below.

3. Are more tickets needed to ride the moon bounce and the water ride or the alpine slide and the roller coaster?

   _____

   _____

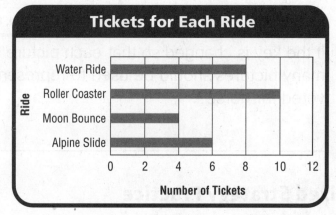

4. If each ticket costs $0.50, how much money does it cost to ride all four rides one time? _____

5. Josie made a bar graph to show how many books her friends have. Which book has the shortest bar?

   A 8 novels

   B 2 sports books

   C 1 math book

   D 5 cookbooks

6. Bertrand made a bar graph to show how many activities were scheduled in the month of May. Which activity has the shortest bar?

   F 2 football games

   G 7 pep rallies

   H 3 tennis matches

   J 9 assemblies

Practice

Make a Bar Graph

For 1–11, use the pictograph below.

Students voted for their favorite cookie. The resulting data is shown in the pictograph.

1.–7. Use the data in the pictograph to make a bar graph below.

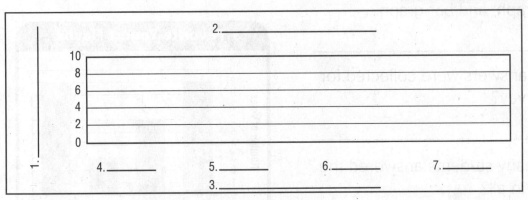

Favorite Cookie

Chocolate Chip	🍪 🍪 🍪 🍪 🍪
Ginger Snaps	🍪 🍪
Oatmeal	🍪 🍪 🍪
Peanut Butter	🍪 🍪 🍪

Key: Each 🍪 = 2 cookies.

2._____

10
8
6
4
2
0

1.

4._____ 5._____ 6._____ 7._____

3._____

Problem Solving and TAKS Prep

8. Which cookie received 1 vote more than ginger snaps but 1 less vote than oatmeal?

9. **Reasoning** If two of the votes for chocolate chip changed to peanut butter, how would the total number of votes change?

10. How many more votes did oatmeal receive than ginger snaps received?

   A 1 vote   **C** 4 votes

   B 2 votes   **D** 8 votes

11. How many more people voted for ginger snaps and oatmeal combined, than for chocolate chip?

   F 1 person   **H** 4 people

   G 2 people   **J** 8 people

Practice

Take a Survey

For 1–4, use the tally table, pictograph, and bar graph below to answer each question.

Favorite Yogurt Flavor	
Flavor	**Tally**
Plain	卌 IIII
Vanilla	卌 I
Cherry	III
Peach	卌 IIII

Favorite Yogurt Flavor

Plain	🍦 🍦 🍦
Vanilla	🍦 🍦
Cherry	🍦
Peach	🍦 🍦 🍦

Key: Each 🍦 = 3 votes.

1. What is the title of the tally table, pictograph, and bar graph?

_____

2. Which answers were collected for the survey?

_____

3. How many students answered the survey in all?

_____

4. Which flavor was chosen the most? Which was chosen the least?

_____

5. Think of a survey question and write it down.

_____

_____

6. Write 4 possible answer choices to your survey question.

_____

_____

7. Set up the tally table at the right to record the results of your survey.

© Harcourt

Classify Data

For 1–5, use the table below.

1. How many small boxes are made with thick cardboard? _____

2. How many thin cardboard boxes are medium? _____

3. How many boxes are large? _____

4. How many more boxes are small than are medium? _____

5. How many boxes are there in all? _____

Boxes		
Size	**Thin Cardboard**	**Thick Cardboard**
Small	5	2
Medium	3	3
Large	2	1

For 6–10, use the geometric figures shown below.

6. Complete the table below to classify the figures at the right.

Geometric Figures		
	White	**Gray**
Circle		
Diamond		

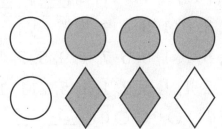

7. How many diamonds are there in all? _____

8. How many figures in all are gray? _____

Problem Solving and TAKS Prep

9. **What if** a third row of 4 red squares were added to the 2 rows of figures above? How would the table look then?

_____

_____

_____

10. **What if** one gray circle were changed to a white diamond? What would the classification of figures be then?

_____

_____

_____

11. Which shows two ways to classify a group of shirts?

   A girl and boy   **C** salty and sweet

   B quiet and loud **D** size and color

12. Which shows one way to classify a group of sports balls?

   F time   **H** sound

   G temperature **J** color

© Harcourt

Patterns

Name a pattern unit. Find the missing number or shape.

1. 2, 5, 9, 2, 5, 9, 2, 5, 9, 2, 5, 9, ☐, 5   _____

2. 5, 0, 9, 0, 5, 0, 9, 0, 5, 0, 9, 0, 5, ☐, 9   _____

3. 6, 1, 9, 2, 6, 1, 9, 2, 6, 1, 9, 2, 6, 1, ☐   _____

4. ☆○□☆○□☆○□☆ _?_ ,   _____

Predict the next two numbers or shapes in each pattern.

5. 3, 7, 3, 7, 3, 7, 3, 7, 3, 7, ☐, ☐

6. 2, 2, 8, 2, 2, 8, 2, 2, 8, 2, 2, 8, ☐, ☐

7. 2, 17, 17, 2, 17, 17, 2, ☐, ☐

8. 0, 1, 3, 0, 1, 3, 0, 1, 3, ☐, ☐

9. 1, 9, 5, 7, 1, 9, 5, 7, 1, 9, 5, 7, 1, 9, 5, 7, ☐, ☐

10. 1, 5, 2, 1, 1, 5, 2, 1, 1, 5, 2, 1, 1, 5, 2, 1, 1, 5, ☐, ☐

11. ○○◖○○◖○○◖○ _?_ _?_   _____

12. ●●●■●●●■●●●■●●● _?_ _?_   _____

Problem Solving and TAKS Prep

13. Alyssa made a bead necklace. Look at the pattern she used.

 _?_ ○◇

Which shape is missing?

_____

14. Phil paints a border on a birdhouse. Look at the pattern he uses.

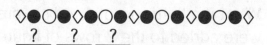
_?_ _?_ _?_

What will the next three shapes be?

_____

15. Which are the next two numbers in the pattern below?

9, 6, 1, 9, 6, 1, 9, 6, 1, ☐, ☐

A 1, 1   **C** 6, 1

B 1, 9   **D** 9, 6

16. Which is the pattern unit in the pattern below?

○○○□○○○□○○○

F □○○□   **H** □○○

G ○○□○   **J** ○○□

Practice

Geometric Patterns

Find the pattern unit or rule. Then name the next figure.

1. △ △△ △ △△△ △△△△

2. ○○○○ ○○○○○○ ○○○○○

_____

_____

_____

_____

3. ▭ ▭ ▭▭ ▭ ▭▭ ▭ ▭▭ ▭

4.

_____

_____

_____

_____

Draw the missing figure.

5. ▢ ▢▢▢ ▢▢▢▢▢ ▢▢▢▢▢▢▢   ___?___   ▢▢▢▢▢▢▢▢▢▢▢

_____

6. ▽ ▲ △ ▽ ▲ △ _?_

_____

Problem Solving and TAKS Prep

7. Sam drew this pattern.

○◯○○ _?_ ○○◯○○

Find the missing figure.

_____

8. Ayla drew this pattern.

○ △ △○ △ △○ △ △

Which figure is next?

_____

9.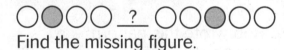

Which figure is next?

10.

Which figure is next?

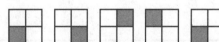

A ⌒⌒⌒⌒⌒

F

H

B ⌒⌒⌒⌒⌒⌒⌒

C ⌒⌒⌒⌒⌒⌒⌒⌒

G

J

D ⌒⌒⌒⌒⌒⌒⌒⌒⌒

Practice

Number Patterns

Write a rule for each pattern. Then find the next number.

1. 15, 21, 27, 33, 39, 45   **2.** 99, 91, 83, 75, 67   **3.** 7, 10, 13, 16, 19, 22

_____   _____   _____

_____   _____   _____

4. 555, 530, 505, 480, 455, 430   **5.** 4, 8, 13, 17, 22, 26, 31, 35, 40, 44

_____   _____

Find the missing numbers.

6. 25, 24, 44, ☐, 63, 62

7. 222, 218, 214, ☐, ☐, 202

8. 27, 44, 61, ☐, 95, ☐, 129

9. 33, 36, 46, 49, 59, ☐, 72, ☐

10. 11, 16, 12, 17, 13, ☐, 14, 19, 15, ☐, 16, ☐, 17

11. 5, 10, 20, 25, 35, ☐, 50, 55, ☐, 70, 80

12. 11, 21, 16, 26, 21, 31, 26, ☐, ☐, 41, ☐, 46

13. 8, 9, 11, 14, 18, 23, 29, 36, ☐, 53, ☐, 74

Problem Solving and TAKS Prep

14. Deanna wrote this pattern:
10, 15, 13, 18, 16, 21, 19, 24.
What rule did Deanna use?

15. Bob wrote this pattern:
17, 34, 51, 68, 85.
What will the next number in
Bob's pattern be?

_____   _____

16. Maria wrote the following pattern.
25, 28, 20, 23, ☐, 18, ☐, 13.
Which numbers are missing?

   A 25, 20   **C** 23, 18

   B 18, 10   **D** 15, 10

17. Eli wrote the following pattern:
12, 23, 22, 33, 32, ☐, ☐, 53.
Which numbers are missing?

   F 43, 42   **H** 42, 52

   G 53, 22   **J** 37, 39

Extend Patterns

Name the rule or pattern unit. Find the next three numbers or shapes.

1. ↑ ↓↓ ↑↑↑ ↓↓↓↓ ↑↑↑↑↑

_____

_____

2. 21, 27, 24, 30, 27, 33, 30, 36, 33

3. 98, 91, 84, 77, 70, 63

_____

4. ☆ ○ ○ ☆ ○ ○ ○ ☆ ○

_____

_____

Find the next three shapes in the pattern.

5. ◗ ◖ ◖◗ ◗ ◖ ◖◗ ◗ ◖ ◖◗ ◗

_____

6. ▲▲ ▽▽▽▽ ▲▲▲▲▲▲ ▽▽▽▽▽▽▽▽

_____

Problem Solving and TAKS Prep

7. Tara wrote a number pattern. She started with the number 9 and used the rule add 6. Write the first five numbers of Tara's pattern.

_____

8. Donald wrote the number pattern below. Write the next three numbers in the pattern.
82, 74, 66, 58, 50, 42, 34, 26

_____

9. Which are the next three shapes in the pattern below?

   A triangle, circle, circle

   B triangle, square, circle

   C circle, circle, square

   D square, circle, circle

10. Carlos made a shape pattern. The pattern unit was star, moon, circle, cloud. Carlos drew 17 shapes and began with star. Which was the 17th shape?

   F moon

   G star

   H cloud

   J circle

Practice

Problem Solving Workshop Strategy: Look for a Pattern

Problem Solving Strategy Practice

Look for a pattern to solve.

1. Max used stamps to make a pattern around the edge of a picture. His pattern unit was 2 triangles, 3 circles, 2 squares. He stamped a total of 28 figures and began with two triangles. Which shape was the 14th figure? _____

2. Al used stamps to make a pattern around the edge of a painting. His pattern unit was 3 triangles, 1 star, 1 square. He stamped a total of 33 figures and began with a star. Which shape was the 33rd figure? _____

3. Kya arranged shape cards to make a pattern. She turned two of the cards face down. Which shapes are on the two cards Kya turned face down?

☆ △ ○ ☆ □ ○ ☆ △ □ _____

Mixed Strategy Practice

USE DATA For exercises 4–5, use the table below.

4. Mara is saving money to buy a new hockey stick. She saved $2 the first week, $4 the second week, $6 the third week, and $8 the fourth week. If this pattern continues, how much will Mara save the fifth week?

Week	Savings
1	$2
2	$4
3	$6
4	$8
5	

_____

5. How much money in all, will Mara have saved during 5 weeks?

_____

6. June has 6 stickers. Arnie has 11 stickers. How many more stickers does Arnie have than June has?

_____

7. Kyle has 4 packages of 6 napkins each. He puts the same number of napkins on each of 8 tables. How many napkins does Kyle put on each table?

_____

© Harcourt

Algebra: Relate Addition to Multiplication

Use counters to model. Then write an addition sentence and a multiplication sentence for each.

1. 3 groups of 5   **2.** 4 groups of 7   **3.** 2 groups of 6   **4.** 4 groups of 6

_____   _____   _____   _____

_____   _____   _____   _____

Write a multiplication sentence for each.

5.    **6.**    **7.**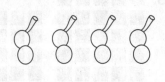

_____   _____   _____

8. 5 + 5 + 5 = 15   **9.** 6 + 6 + 6 = 18   **10.** 7 + 7 + 7 = 21

_____   _____   _____

11. 3 + 3 + 3 + 3 = 12   **12.** 8 + 8 + 8 = 24   **13.** 5 + 5 + 5 + 5 = 20

_____   _____   _____

Problem Solving and TAKS Prep

14. Mike is baking apple bread. He uses 2 apples for every loaf of bread. He makes 4 loaves of bread. How many apples does Mike use in all?

_____

15. Cynthia is making small pizzas. She puts 4 mushrooms on each pizza. How many mushrooms does Cynthia use to make 3 pizzas?

_____

16. Which is another way to show 3 + 3 + 3 + 3?

A 4×3

B 4×4

C 3×12

D 3×3

17. Which is another way to show 6 + 6 + 6?

F 6×4

G 3×3

H 3×6

J 6×6

Practice

Model with Arrays

Write a multiplication sentence for each array.

1.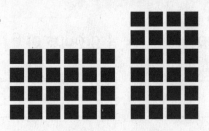

_____

2.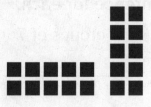

_____

3.

_____

4.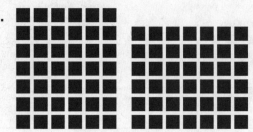

_____

Problem Solving and TAKS Prep

5. Jerry put 30 cans of tomatoes in 6 rows. How many cans were in each row?

_____

6. Maya pulled 6 carrots each from 2 rows in her garden. She used 4 carrots to make soup. How many carrots, from these that she pulled, does Maya have left?

_____

7. Kayla planted carrot seeds in 5 rows. She planted 9 seeds in each row. Which number sentence shows how many seeds Kayla planted?

  A $9 + 5 = 14$ **C** $5 \times 5 = 25$

  B $5 \times 9 = 45$ **D** $9 \times 9 = 81$

8. Chet stacked blocks to make a wall. He used 32 blocks. He put 8 blocks in each row. How many rows did Chet make?

  F 4   **H** 9

  G 6   **J** 12

© Harcourt

Name_____

Lesson 9.3

Multiply with 2

Write a multiplication sentence for each.

1.   **2.**    **3.**   **4.**

_____   _____   _____   _____

Find the product.

5. $2 \times 7 =$ ____   **6.** $5 \times 2 =$ ____   **7.** $2 \times 4 =$ ____   **8.** $3 \times 2 =$ ____

9. $\begin{array}{r} 2 \\ \times 3 \\ \hline \end{array}$ **10.** $\begin{array}{r} 5 \\ \times 2 \\ \hline \end{array}$ **11.** $\begin{array}{r} 2 \\ \times 8 \\ \hline \end{array}$ **12.** $\begin{array}{r} 2 \\ \times 6 \\ \hline \end{array}$ **13.** $\begin{array}{r} 3 \\ \times 2 \\ \hline \end{array}$ **14.** $\begin{array}{r} 2 \\ \times 4 \\ \hline \end{array}$

15. $\begin{array}{r} 6 \\ \times 2 \\ \hline \end{array}$ **16.** $\begin{array}{r} 2 \\ \times 7 \\ \hline \end{array}$ **17.** $\begin{array}{r} 2 \\ \times 2 \\ \hline \end{array}$ **18.** $\begin{array}{r} 7 \\ \times 2 \\ \hline \end{array}$ **19.** $\begin{array}{r} 4 \\ \times 2 \\ \hline \end{array}$ **20.** $\begin{array}{r} 9 \\ \times 2 \\ \hline \end{array}$

Problem Solving and TAKS Prep

21. Seven friends go for a swim. Each pays $2 to use the town pool. How much money do the friends pay in all, to use the pool?

_____

22. Darius and Marvin each wear 3 costumes in the school play. How many costumes do Darius and Marvin wear in all?

_____

23. Savannah and George each wore 4 costumes in the school play. Which number sentence shows Savannah and George's total number of costumes worn?

A $2 \times 4 = 8$

B $3 \times 2 = 6$

C $5 + 2 = 7$

D $4 \times 2 = 6$

24. There are 2 rows, with 9 cans in each row. Which number sentence shows how many cans there are in all?

F $9 + 2 = 11$

G $9 \times 3 = 21$

H $2 + 9 = 11$

J $2 \times 9 = 18$

© Harcourt

PW53

Practice

Multiply with 4

Find the product.

1. ●●●●
   ●●●●
   ●●●●
   ●●●●
   ●●●●
   ●●●●
   ●●●●
   ●●●●

2. ●●●●●●●
   ●●●●●●●
   ●●●●●●●
   ●●●●●●●

3. ●●●●●●●●●●
   ●●●●●●●●●●
   ●●●●●●●●●●
   ●●●●●●●●●●

_____   _____   _____

4. $4 \times 5 =$ ___

5. $4 \times 4 =$ ___

6. $2 \times 4 =$ ___

7. $3 \times 4 =$ ___

8. $\begin{array}{r} 4 \\ \times 3 \\ \hline \end{array}$

9. $\begin{array}{r} 5 \\ \times 4 \\ \hline \end{array}$

10. $\begin{array}{r} 4 \\ \times 6 \\ \hline \end{array}$

11. $\begin{array}{r} 4 \\ \times 9 \\ \hline \end{array}$

12. $\begin{array}{r} 7 \\ \times 4 \\ \hline \end{array}$

13. $\begin{array}{r} 8 \\ \times 4 \\ \hline \end{array}$

Problem Solving and TAKS Prep

14. Mary's brother gave her some toy cars. These toy cars have 36 wheels in all. Each car has 4 wheels. How many toy cars did Mary receive?

15. Eli has 3 toy cars. Andy has 2 toy cars. Amanda has 4 toy cars. Each toy car has 4 wheels. How many wheels do their toy cars have in all?

16. Sasha has 7 toy cars. Each toy car has 4 wheels. How many wheels do Sasha's toy cars have in all?

   A 11   C 24

   B 21   D 28

17. There are 4 rows of 8 toy cars on a shelf. Which number sentence shows how many toy cars there are on the shelf in all?

   F $8 + 4 = 12$ H $4 \times 8 = 32$

   G $9 \times 4 = 36$ J $4 \times 7 = 28$

Practice

Algebra: Multiply with 1 and 0

Find the product.

1. $6 \times 1 =$ ___ **2.** $0 \times 9 =$ ___ **3.** $1 \times 4 =$ ___ **4.** $8 \times 0 =$ ___

5. $\begin{array}{r} 0 \\ \times 6 \\ \hline \end{array}$ **6.** $\begin{array}{r} 9 \\ \times 1 \\ \hline \end{array}$ **7.** $\begin{array}{r} 4 \\ \times 0 \\ \hline \end{array}$ **8.** $\begin{array}{r} 5 \\ \times 1 \\ \hline \end{array}$ **9.** $\begin{array}{r} 3 \\ \times 0 \\ \hline \end{array}$ **10.** $\begin{array}{r} 1 \\ \times 8 \\ \hline \end{array}$

11. $\begin{array}{r} 2 \\ \times 1 \\ \hline \end{array}$ **12.** $\begin{array}{r} 1 \\ \times 6 \\ \hline \end{array}$ **13.** $\begin{array}{r} 1 \\ \times 4 \\ \hline \end{array}$ **14.** $\begin{array}{r} 0 \\ \times 1 \\ \hline \end{array}$ **15.** $\begin{array}{r} 3 \\ \times 1 \\ \hline \end{array}$ **16.** $\begin{array}{r} 1 \\ \times 0 \\ \hline \end{array}$

Write a multiplication sentence shown on each number line.

17.

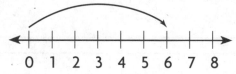

_____

18.

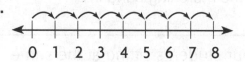

_____

Find the missing number.

19. $5 \times \boxed{} = 0$ **20.** $1 \times \boxed{} = 9$ **21.** $7 \times \boxed{} = 7$ **22.** $0 \times 52 =$ ___

Problem Solving and TAKS Prep

23. At a farm, Kaitlyn saw 9 rabbits. Each rabbit was near its own water bowl. How many water bowls did Kaitlyn see at the farm?

_____

24. Cody saw 8 calves on his visit to a farm. None of the calves had horns. How many horns did Cody see at the farm?

_____

25. Chloe has 6 pockets. Each pocket contains 1 coin. Which number sentence shows how many coins Chloe has in all of her pockets combined?

A $1 + 6 = 7$ **C** $6 \times 1 = 6$

B $0 \times 6 = 0$ **D** $6 \times 0 = 6$

26. Len has 7 pockets. He has 0 coins in each pocket. Which number sentence shows how many coins Len has in all of his pockets combined?

F $7 \times 0 = 7$ **H** $7 \times 1 = 7$

G $0 \times 7 = 0$ **J** $1 + 7 = 8$

Problem Solving Workshop Strategy: Draw a Picture

Problem Solving Strategy Practice

Draw a picture to solve.

1. Mr. Jardin has 8 tomato plants. On each tomato plant there are 7 ripe tomatoes. How many ripe tomatoes does Mr. Jardin have in all?

   _____

2. In a marching band, there are 4 rows of horn players. Each row has 9 horn players. How many horn players are in the marching band in all?

   _____

3. Four students sitting at the same table have apple slices in their lunches. If each student has 6 slices, how many apple slices are at the table in all?

   _____

Mixed Strategy Practice

4. There are 8 drummers in a marching band. Each drummer has 2 drum sticks. How many drum sticks do the drummers have in all?

   _____

5. Matthew is making a large pizza for his party. There are 8 people at the party. Each person will eat 1 slice. How many slices should Matthew cut the pizza into?

   _____

6. At Adam's lunch table, 7 students have a serving of peas and no one has a serving of spinach. How many servings of peas and spinach are at Adam's lunch table in all? Show your work.

   _____

7. **Open-Ended** Lea makes a necklace with 5 beads. She strings a red bead first and last. The second and fourth beads are yellow. The middle bead is orange. Describe the pattern of beads.

   _____

Practice

© Harcourt

Multiply with 5 and 10

Find the product.

1. $10 \times 7 =$ ____ 2. ____ $= 5 \times 4$ 3. $8 \times 10 =$ ____ 4. ____ $= 5 \times 7$

5. $0 \times 10 =$ ____ 6. ____ $= 10 \times 4$ 7. $5 \times 1 =$ ____ 8. ____ $= 10 \times 3$

9. $2 \times 5 =$ ____ 10. $0 \times 10 =$ ____ 11. $10 \times 8 =$ ____ 12. ____ $= 5 \times 3$

13. $\begin{array}{r} 3 \\ \times 5 \\ \hline \end{array}$
14. $\begin{array}{r} 10 \\ \times 1 \\ \hline \end{array}$
15. $\begin{array}{r} 5 \\ \times 5 \\ \hline \end{array}$
16. $\begin{array}{r} 4 \\ \times 5 \\ \hline \end{array}$
17. $\begin{array}{r} 5 \\ \times 10 \\ \hline \end{array}$
18. $\begin{array}{r} 10 \\ \times 7 \\ \hline \end{array}$

19. $\begin{array}{r} 10 \\ \times 4 \\ \hline \end{array}$
20. $\begin{array}{r} 9 \\ \times 5 \\ \hline \end{array}$
21. $\begin{array}{r} 7 \\ \times 5 \\ \hline \end{array}$
22. $\begin{array}{r} 5 \\ \times 1 \\ \hline \end{array}$
23. $\begin{array}{r} 5 \\ \times 6 \\ \hline \end{array}$
24. $\begin{array}{r} 10 \\ \times 9 \\ \hline \end{array}$

Problem Solving and TAKS Prep

25. A car can carry up to 5 people. There are 6 cars. What is the greatest number of people who can ride in these cars at one time?

26. The school choir is standing in 6 rows with 10 students in each row. How many students are in the school choir?

_____

27. A tableware setting includes 5 pieces: 2 spoons, 2 forks, and 1 knife. How many pieces are included in 8 tableware settings?

A 13
B 20
C 40
D 80

28. A doctor can help 10 patients each day. If an office has 5 doctors, what is the greatest number of patients they can help each day?

F 15
G 50
H 100
J 150

Multiply with 3

Find the product.

1. $4 \times 3 =$ _____

2. $7 \times 3 =$ _____

3. _____ $= 3 \times 9$

4. _____ $= 5 \times 3$

5. _____ $= 3 \times 3$

6. $5 \times 3 =$ _____

7. _____ $= 3 \times 8$

8. $6 \times 3 =$ _____

9. _____ $= 3 \times 0$

10. $\begin{array}{r} 6 \\ \times 3 \\ \hline \end{array}$

11. $\begin{array}{r} 3 \\ \times 1 \\ \hline \end{array}$

12. $\begin{array}{r} 4 \\ \times 3 \\ \hline \end{array}$

13. $\begin{array}{r} 8 \\ \times 3 \\ \hline \end{array}$

14. $\begin{array}{r} 7 \\ \times 3 \\ \hline \end{array}$

15. $\begin{array}{r} 9 \\ \times 3 \\ \hline \end{array}$

16. $\begin{array}{r} 0 \\ \times 3 \\ \hline \end{array}$

17. $\begin{array}{r} 3 \\ \times 3 \\ \hline \end{array}$

Problem Solving and TAKS Prep

18. A design contains 5 triangles. How many sides do 5 triangles include?

19. A boat can carry up to 3 people. What is the minimum number of boats needed to carry 24 people? Explain.

20. There are 8 buns in each bag of hamburger buns. If you have 3 bags of hamburger buns, how many hamburger buns do you have in all?

    A 8

    B 11

    C 16

    D 24

21. A pint of ice cream serves 3 people. How many people are served by 5 pints of ice cream?

    F 3

    G 5

    H 15

    J 30

© Harcourt

Practice

Multiply with 6

Find the product.

1. $9 \times 6 =$ ___

2. ___ $= 6 \times 8$

3. $4 \times 6 =$ ___

4. ___ $= 6 \times 7$

5. $6 \times 1 =$ ___

6. ___ $= 6 \times 6$

7. $6 \times 0 =$ ___

8. ___ $= 5 \times 6$

9. $5 \times 5 =$ ___

10.
$$\begin{array}{r} 4 \\ \times 6 \\ \hline \end{array}$$

11.
$$\begin{array}{r} 9 \\ \times 6 \\ \hline \end{array}$$

12.
$$\begin{array}{r} 6 \\ \times 8 \\ \hline \end{array}$$

13.
$$\begin{array}{r} 6 \\ \times 1 \\ \hline \end{array}$$

14.
$$\begin{array}{r} 6 \\ \times 9 \\ \hline \end{array}$$

15.
$$\begin{array}{r} 6 \\ \times 7 \\ \hline \end{array}$$

16.
$$\begin{array}{r} 2 \\ \times 6 \\ \hline \end{array}$$

17.
$$\begin{array}{r} 6 \\ \times 6 \\ \hline \end{array}$$

Problem Solving and TAKS Prep

18. A lecture room contains 9 rows, with 6 chairs in each row. How many chairs are in the lecture room?

19. Lila saw 6 ducks. Each duck has 2 wings. How many wings do the 6 ducks have?

20. Ken has 6 pages of stickers. Each page contains 8 stickers. How many stickers does Ken have?

A 40
B 46
C 48
D 60

21. Heavy-duty pickup trucks hold 6 tires. How many tires do 5 heavy-duty pickup trucks hold?

F 30
G 36
H 55
J 60

Practice

Algebra: Practice the Facts

Find the product.

1. $10 \times 8 =$ ____

2. $3 \times 0 =$ ____

3. ____ $= 4 \times 6$

4. ____ $= 9 \times 3$

5. $6 \times 5 =$ ____

6. ____ $= 2 \times 8$

7. ____ $= 1 \times 5$

8. $6 \times 10 =$ ____

9. $5 \times 3 =$ ____

10. $\begin{array}{r} 3 \\ \times 4 \\ \hline \end{array}$

11. $\begin{array}{r} 6 \\ \times 6 \\ \hline \end{array}$

12. $\begin{array}{r} 9 \\ \times 1 \\ \hline \end{array}$

13. $\begin{array}{r} 7 \\ \times 5 \\ \hline \end{array}$

Show two different ways to find each product.

14. $3 \times 7 =$ ____

15. ____ $= 5 \times 2$

Problem Solving and TAKS Prep

16. A cow eats 2 bales of hay in one week. How many bales of hay does a cow eat in 6 weeks?

_____

17. Ryan has 21 baseballs. If he keeps them in 3 even rows, how many baseballs are in each row?

_____

18. Which multiplication fact does the picture below show?

A $5 \times 3 = 15$ **C** $5 \times 5 = 25$

B $4 \times 5 = 20$ **D** $6 \times 5 = 30$

19. Glenn bought 5 packages of postcards. Each package included 10 postcards. How many postcards did Glenn buy? Explain.

_____

_____

_____

Practice

Problem Solving Workshop Strategy: Act It Out

Problem Solving Strategy Practice

Act out the problem to solve.

1. Luis puts ice cubes into glasses for his friends' drinks. He puts 3 ice cubes into each glass. How many ice cubes does Luis use if he has 9 friends?

   _____

2. Rebecca hands out coupons. She gives 4 coupons to each customer. How many coupons does Rebecca hand out if she has 6 customers?

   _____

3. Four men are in a line. Fred is in front of Rex. Ken is behind Rex. William is in front of Fred. Who is first in line?

   _____

4. Vic is handing out pencils for drawing. Each student receives 5 pencils. How many pencils does Vic hand out if there are 9 students?

   _____

Mixed Strategy Practice

5. Donald rolls sushi. It takes him 5 minutes to make each roll. How many minutes would it take Donald to make 7 rolls?

   _____

6. Tina has 4 dimes, 5 nickels, and 4 pennies. How much money does Tina have in all?

   _____

USE DATA For 7–8, use the table below.

7. Jenny bought 3 packages of T-shirts. How many T-shirts did she buy in all?

   _____

8. Which contains more items, 3 packages of socks or 3 packages of headbands?

   _____

Clothing Packages	
Item	Number in Package
Socks	6
T-shirts	2
Headbands	4

Multiply with 8

Find the product.

1. $8 \times 3 =$ _____   **2.** $10 \times 8 =$ _____   **3.** $1 \times 8 =$ _____   **4.** $7 \times 5 =$ _____

5. $7 \times 9 =$ _____   **6.** $8 \times 4 =$ _____   **7.** $8 \times 9 =$ _____   **8.** $4 \times 4 =$ _____

9. $\begin{array}{r} 8 \\ \times 7 \\ \hline \end{array}$
10. $\begin{array}{r} 1 \\ \times 8 \\ \hline \end{array}$
11. $\begin{array}{r} 3 \\ \times 7 \\ \hline \end{array}$
12. $\begin{array}{r} 3 \\ \times 8 \\ \hline \end{array}$
13. $\begin{array}{r} 6 \\ \times 3 \\ \hline \end{array}$
14. $\begin{array}{r} 9 \\ \times 8 \\ \hline \end{array}$

15. $\begin{array}{r} 6 \\ \times 8 \\ \hline \end{array}$
16. $\begin{array}{r} 4 \\ \times 8 \\ \hline \end{array}$
17. $\begin{array}{r} 2 \\ \times 9 \\ \hline \end{array}$
18. $\begin{array}{r} 8 \\ \times 2 \\ \hline \end{array}$
19. $\begin{array}{r} 8 \\ \times 8 \\ \hline \end{array}$
20. $\begin{array}{r} 5 \\ \times 8 \\ \hline \end{array}$

Problem Solving and TAKS Prep

USE DATA For 21–22, use the table below.

21. If Kaylie's beanstalk grows the same amount every week, how tall will it be after 6 weeks?

_____

22. If the beanstalks grow the same amount each week, how much taller than Amy's beanstalk will Kaylie's beanstalk be, after 8 weeks?

_____

Growth of Beanstalks in 1 Week	
Student	Height of Beanstalk
Kaylie	8 inches
Bret	6 inches
Amy	4 inches

23. At the dog park, there are 8 dogs. Each dog is given 3 bones. How many bones are given out at the dog park?

A 21   **C** 23

B 24   **D** 28

24. There are 6 pieces of fruit in each bag. Sandra buys 8 bags. How many pieces of fruit does Sandra buy?

F 42   **H** 45

G 48   **J** 14

Algebra: Patterns with 9

Find each product.

1. _____ $= 9 \times 3$ 2. $9 \times 4 =$ _____   3. _____ $= 9 \times 8$ 4. $9 \times 5 =$ _____

5. $7 \times 9 =$ _____   6. _____ $= 3 \times 4$ 7. $9 \times 9 =$ _____   8. _____ $= 5 \times 4$

9. $\begin{array}{r} 9 \\ \times 1 \\ \hline \end{array}$ 10. $\begin{array}{r} 9 \\ \times 2 \\ \hline \end{array}$ 11. $\begin{array}{r} 6 \\ \times 3 \\ \hline \end{array}$ 12. $\begin{array}{r} 9 \\ \times 6 \\ \hline \end{array}$ 13. $\begin{array}{r} 9 \\ \times 7 \\ \hline \end{array}$ 14. $\begin{array}{r} 9 \\ \times 8 \\ \hline \end{array}$

Compare. Write $<$, $>$, or $=$ for each $\bigcirc$.

15. $5 \times 8 \bigcirc 6 \times 7$ 16. $9 \times 3 \bigcirc 4 \times 7$ 17. $3 \times 6 \bigcirc 2 \times 8$

18. $4 \times 3 \bigcirc 2 \times 6$ 19. $9 \times 4 \bigcirc 6 \times 6$ 20. $9 \times 5 \bigcirc 8 \times 4$

Problem Solving and TAKS Prep

21. A model of the solar system includes 8 planets. How many planets are in 8 models?

_____

22. Bob has 4 plants. Ron has 9 times as many plants as Bob has. How many plants does Ron have?

_____

23. A package of pencils contains 9 pencils. How many pencils are in 3 packages?

   A 6
   B 9
   C 18
   D 27

24. Ms. Lee took 9 children to the zoo. Each ticket cost $4. How much did it cost for the 9 children to go to the zoo?

   F $4
   G $9
   H $13
   J $36

Practice

Multiply with 7

Find the product.

1. $7 \times 3 =$ _____ **2.** $9 \times 7 =$ _____ **3.** $7 \times 8 =$ _____ **4.** $6 \times 5 =$ _____

5. $7 \times 1 =$ _____ **6.** $4 \times 7 =$ _____ **7.** $6 \times 8 =$ _____ **8.** $5 \times 7 =$ _____

9. $\begin{array}{r} 8 \\ \times 5 \\ \hline \end{array}$ **10.** $\begin{array}{r} 2 \\ \times 7 \\ \hline \end{array}$ **11.** $\begin{array}{r} 6 \\ \times 7 \\ \hline \end{array}$ **12.** $\begin{array}{r} 7 \\ \times 7 \\ \hline \end{array}$ **13.** $\begin{array}{r} 9 \\ \times 7 \\ \hline \end{array}$ **14.** $\begin{array}{r} 7 \\ \times 5 \\ \hline \end{array}$

15. $\begin{array}{r} 4 \\ \times 6 \\ \hline \end{array}$ **16.** $\begin{array}{r} 7 \\ \times 4 \\ \hline \end{array}$ **17.** $\begin{array}{r} 8 \\ \times 7 \\ \hline \end{array}$ **18.** $\begin{array}{r} 9 \\ \times 3 \\ \hline \end{array}$ **19.** $\begin{array}{r} 7 \\ \times 1 \\ \hline \end{array}$ **20.** $\begin{array}{r} 7 \\ \times 6 \\ \hline \end{array}$

Problem Solving and TAKS Prep

USE DATA For 21–22, use the table below.

21. Molly is going to make snack mix for Ben's party. She wants to make 7 batches. How many cups of wheat cereal will Molly need?

Snack Mix Recipe for 1 Batch	
Snack	**Number of Cups**
Wheat Cereal	4
Rice Crisps	2
Sesame Toasts	1

_____

22. If Molly makes 7 batches of snack mix, how many cups of snacks will she need in all? _____

23. Adriana is making muffins with a mold that holds 7 muffins. How many muffins can Adriana make with 4 molds?

   A 14

   B 21

   C 28

   D 35

24. A box holds 7 dog biscuits. Dan has 3 boxes of dog biscuits. How many dog biscuits does Dan have?

   F 14

   G 21

   H 28

   J 35

Practice

Algebra: Practice the Facts

Find the product.

1. $4 \times 5 =$ _____   2. _____ $= 8 \times 9$ 3. $7 \times 5 =$ _____   4. _____ $= 6 \times 6$

5. $3 \times 2 =$ _____   6. _____ $= 6 \times 7$ 7. _____ $= 9 \times 4$ 8. $5 \times 8 =$ _____

9. $\begin{array}{r} 4 \\ \times 9 \\ \hline \end{array}$ 10. $\begin{array}{r} 2 \\ \times 7 \\ \hline \end{array}$ 11. $\begin{array}{r} 8 \\ \times 8 \\ \hline \end{array}$ 12. $\begin{array}{r} 9 \\ \times 3 \\ \hline \end{array}$ 13. $\begin{array}{r} 5 \\ \times 8 \\ \hline \end{array}$ 14. $\begin{array}{r} 7 \\ \times 6 \\ \hline \end{array}$

Find the missing number.

15. $\boxed{} \times 8 = 32$ 16. $7 \times 8 = \boxed{}$ 17. $\boxed{} \times 6 = 24$

18. $5 \times \boxed{} = 45$ 19. $\boxed{} \times 9 = 27$ 20. $6 \times \boxed{} = 48$

Explain two different ways to find the product.

21. _____ $= 9 \times 9$ _____

22. _____ $= 10 \times 8$ _____

Compare. Write <, >, or = for each $\bigcirc$.

23. $3 \times 8 \bigcirc 4 \times 6$ 24. $9 \times 5 \bigcirc 6 \times 8$ 25. $4 \times 7 \bigcirc 9 \times 3$

Problem Solving and TAKS Prep

26. Each basketball team has 8 players. How many players are on 7 basketball teams?

   _____

27. Each tennis team has 9 players. How many players are on 3 tennis teams?

   _____

28. Which is the correct number sentence for the array?

   A $7 \times 6 = 42$
   B $6 \times 8 = 48$
   C $6 \times 7 = 48$
   D $8 \times 6 = 42$

29. Which is greater than 9×4?

   F 3×9
   G 5×7
   H 8×5
   J 5×6

Practice

Multiply with 11 and 12

Find the product.

1. $4 \times 11 =$____   **2.** $12 \times 3 =$____   **3.** $7 \times 10 =$____   **4.** ____$= 12 \times 8$

5. $11 \times 0 =$____   **6.** ____$= 5 \times 7$ **7.** $12 \times 7 =$____   **8.** $9 \times 10 =$____

9. $\begin{array}{r} 7 \\ \times 6 \\ \hline \end{array}$ **10.** $\begin{array}{r} 10 \\ \times 5 \\ \hline \end{array}$ **11.** $\begin{array}{r} 12 \\ \times 6 \\ \hline \end{array}$ **12.** $\begin{array}{r} 11 \\ \times 7 \\ \hline \end{array}$ **13.** $\begin{array}{r} 12 \\ \times 5 \\ \hline \end{array}$ **14.** $\begin{array}{r} 11 \\ \times 9 \\ \hline \end{array}$

Problem Solving and TAKS Prep

USE DATA For 15–16, use the graph.

15. The graph shows the number of miles some students live from school. How many miles will Zack travel to and from school in 11 school days?

_____

16. How many miles will Carolyn travel to and from school in 12 school days?

_____

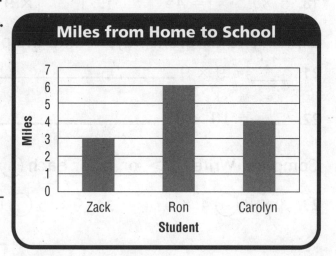

Miles from Home to School

17. Which is the product?

$5 \times 11 =$ ____

A 50

B 55

C 60

D 65

18. A carton of eggs holds 12 eggs. How many eggs are in 5 cartons?

F 50

G 65

H 60

J 55

Problem Solving Workshop Strategy: Compare Strategies

Problem Solving Strategy Practice

1. Bobcats can have a litter of 3 kittens.
   What is the greatest number of kittens that 7 bobcats could have?

   _____

Draw a picture to solve.

Make a table to solve.

2. June goes hiking 4 times each week.
   How many times does June go hiking in 6 weeks?

   _____

Draw a picture to solve.

Make a table to solve.

Mixed Strategy Practice

3. William always sees 8 bats in his backyard at sundown. How many bats does William see in 5 days? Show your work.

   _____

4. **USE DATA** How many students voted for a favorite drink in all? Show your work.

   _____

Favorite Drinks	
Drink	**Number of Votes**
Orange Juice	8
Milk	5
Water	3

Find a Rule

Write a rule for each table. Then complete the table.

1. _____

Children	1	2	3	4	5
Number of Backpacks	5	10			

2. _____

Games	2	3	4	5	6
Players	6	9			

3. _____

Maps	1	2	3	4	5
Cost	$4	$8			

4. _____

Maps	3	4	5	6	7
Campers	6	8			

Problem Solving and TAKS Prep

USE DATA For 5–6, use the table below.

5. Write a rule for the information in this table.

_____

Canoes	1	2	3	4
Campers	3	6	9	

6. How many campers can fit into 4 canoes? _____

7. One rowboat holds 6 people. How many people can 5 rowboats hold?

    A 15   **C** 30

    B 16   **D** 36

8. Each camper needs 2 graham crackers to make s'mores. How many graham crackers do 5 campers need to make s'mores?

    F 10   **H** 25

    G 20   **J** 50

© Harcourt

Practice

Missing Factors

Find the missing factor.

1. $\boxed{} \times 5 = 30$
2. $\boxed{} \times 7 = 28$
3. $4 \times \boxed{} = 16$

4. $\boxed{} \times 9 = 27$
5. $9 \times \boxed{} = 36$
6. $\boxed{} \times 8 = 56$

7. $5 \times \boxed{} = 40$
8. $6 \times \boxed{} = 48$
9. $\boxed{} \times 3 = 18$

10. $n \times 7 = 56$
11. $5 \times k = 45$
12. $3 \times g = 12$

_____   _____   _____

13. $d \times 5 = 10 + 5$
14. $4 \times t = 8 \times 3$
15. $a \times 7 = 30 - 2$

_____   _____   _____

Problem Solving and TAKS Prep

16. Chloe went camping. She brought enough food for 18 meals. She ate 3 meals a day. How many days' worth of food did Chloe bring?

17. Lisa is having a cookout. She wants to make 18 hot dogs. The hot dog buns she is buying come in packages of 6. How many packages of hot dog buns does Lisa need to buy?

_____

18. Which is the missing factor?

$$\boxed{} \times 4 = 36$$

19. Todd wants to bring juice to a picnic. There will be 24 people at the picnic. The juice comes in packages of 6. How many packages will Todd need to bring so that each person receives one juice?

A 6
B 7
C 8
D 9

F 3
G 4
H 6
J 8

Practice

Multiply 3 Factors

Find the product.

1. $(4 \times 2) \times 3$ **2.** $7 \times (2 \times 4)$ **3.** $(5 \times 1) \times 9$ **4.** $(3 \times 3) \times 2$

_____   _____   _____   _____

5. $6 \times (2 \times 2)$ **6.** $(4 \times 1) \times 4$ **7.** $(2 \times 3) \times 6$ **8.** $7 \times (2 \times 2)$

_____   _____   _____   _____

Use parentheses. Find the product.

9. $2 \times 3 \times 5$ **10.** $1 \times 7 \times 6$ **11.** $3 \times 2 \times 6$ **12.** $4 \times 2 \times 7$

_____   _____   _____   _____

13. $3 \times 3 \times 9$ **14.** $6 \times 4 \times 2$ **15.** $7 \times 8 \times 1$ **16.** $9 \times 3 \times 2$

_____   _____   _____   _____

Find the missing factor.

17. $(3 \times \boxed{}) \times 5 = 30$ **18.** $7 \times (\boxed{} \times 2) = 42$ **19.** $(\boxed{} \times 4) \times 6 = 48$

Problem Solving and TAKS Prep

20. A roller coaster contains 2 trains. Each train contains 10 rows of seats. Each row contains 2 seats. How many seats are on the roller coaster?

_____

21. A roller coaster contains 5 cars. Each car contains 2 rows of seats. Each row contains 2 seats. How many seats are on the roller coaster?

_____

22. Which is the answer.

$4 \times 5 \times 2 = \boxed{}$.

   A 18
   B 20
   C 40
   D 50

23. A subway train contains 2 cabs. Each cab contains 5 rows. Each row contains 5 seats. How many seats are on the subway train?

   F 40
   G 50
   H 60
   J 70

         Practice

Multiplication Properties

Find the product. Tell which property you used.

1. 4×3

2. 1×9

3. 7×0

_____   _____   _____

4. $(2 \times 3) \times 6$

5. 4×9

6. $2 \times (3 \times 3)$

_____   _____   _____

7. 8×1

8. 7×3

9. 0×5

_____   _____   _____

10. 6×7

11. $4 \times (5 \times 1)$

12. 6×3

_____   _____   _____

Find the missing factor.

13. $6 \times \boxed{} = 8 \times 6$ **14.** $7 \times 0 = \boxed{} \times 7$ **15.** $(2 \times \boxed{}) \times 7 = 2 \times (2 \times 7)$

Problem Solving and TAKS Prep

16. Holly bought 4 balls of yarn. Each ball of yarn cost $7. How much money did Holly spend?

17. Alice wants to knit 3 hats. She needs 2 balls of yarn for each hat. How many balls of yarn will Alice use?

_____   _____

18. Which is an example of the Zero Property of Multiplication?

    A $2 \times 1 = 2$

    B $2 \times 7 = 7 \times 2$

    C $2 \times 0 = 0$

    D $2 \times 7 (2 \times 4) = (2 \times 2) \times 4$

19. Which is an example of the Associative Property of Multiplication?

    F $4 \times 6 = 6 \times 4$

    G $(2 \times 2) \times 5 = 2 \times (2 \times 5)$

    H $0 \times 7 = 0$

    J $8 \times 1 = 8$

Practice

Problem Solving Workshop Skill: Multistep Problems

Problem Solving Skill Practice

1. Tickets for a movie cost $8 for adults and $6 for children. The Kim family buys 5 tickets. They buy 2 adult tickets and 3 child tickets. How much does it cost for the Kim family to go to the movie?

_____

2. A summer camp rented 2 canoes and 3 paddleboats. Each canoe holds 3 people and each paddleboat holds 4 people. How many people, at one time, can go out on the canoes and paddleboats the summer rented?

_____

3. Stan is at the circus. He buys 4 drinks and 2 sandwiches. The drinks cost $3 each, and the sandwiches cost $4 each. How much does Stan spend in all?

_____

Mixed Applications

4. **USE DATA** Jane went shopping for school supplies. She bought 2 packages of pens and 3 erasers. How much did Jane spend in all?

School Supplies	
Item	Cost
Pens	$3 per package
Markers	$6 per package
Erasers	$1 each
Folders	50¢ each

_____

5. David received a bicycle as a present. He rode his bike 7 miles the first week he had it and 10 miles the second week he had it. During the third week he had it, David rode his bike twice as many miles as he had ridden it the first two weeks combined. How many miles did David ride his bike during the third week that he had it?

_____

Practice

Multiples on a Hundred Chart

Find and list the missing multiples in each chart.

1. Multiples of 10.

_____

1	2	3	4	5	6	7	8	9	10
11	12	13	14	15	16	17	18	19	20
21	22	23	24	25	26	27	28	29	30
31	32	33	34	35	36	37	38	39	40
41	42	43	44	45	46	47	48	49	50
51	52	53	54	55	56	57	58	59	60
61	62	63	64	65	66	67	68	69	70

2. Multiples of 3.

_____

1	2	3	4	5	6	7	8	9	10
11	12	13	14	15	16	17	18	19	20
21	22	23	24	25	26	27	28	29	30
31	32	33	34	35	36	37	38	39	40
41	42	43	44	45	46	47	48	49	50
51	52	53	54	55	56	57	58	59	60
61	62	63	64	65	66	67	68	69	70

3. Multiples of 9.

_____

21	22	23	24	25	26	27	28	29	30
31	32	33	34	35	36	37	38	39	40
41	42	43	44	45	46	47	48	49	50
51	52	53	54	55	56	57	58	59	60
61	62	63	64	65	66	67	68	69	70
71	72	73	74	75	76	77	78	79	80
81	82	83	84	85	86	87	88	69	90

4. Multiples of 4.

11	12	13	14	15	16	17	18	19	20
21	22	23	24	25	26	27	28	29	30
31	32	33	34	35	36	37	38	39	40
41	42	43	44	45	46	47	48	49	50
51	52	53	54	55	56	57	58	59	60
61	62	63	64	65	66	67	68	69	70
71	72	73	74	75	76	77	78	79	80

5. Multiples of 7.

1	2	3	4	5	6	7	8	9	10
11	12	13	14	15	16	17	18	19	20
21	22	23	24	25	26	27	28	29	30
31	32	33	34	35	36	37	38	39	40
41	42	43	44	45	46	47	48	49	50
51	52	53	54	55	56	57	58	59	60
61	62	63	64	65	66	67	68	69	70

6. Multiples of 6.

11	12	13	14	15	16	17	18	19	20
21	22	23	24	25	26	27	28	29	30
31	32	33	34	35	36	37	38	39	40
41	42	43	44	45	46	47	48	49	50
51	52	53	54	55	56	57	58	59	60
61	62	63	64	65	66	67	68	69	70
71	72	73	74	75	76	77	78	79	80

7. In the 9s chart, which number would be shaded after 90?_____

8. In the 6s chart, which number would be shaded after 78?_____

9. In the 10s chart, which number would be shaded after 70?_____

10. In the 7s chart, which number would be shaded after 70?_____

11. In the 3s chart, which number would be shaded after 69?_____

12. In the 4s chart, which number would be shaded after 80?_____

13. Look at the 3s chart and the 7s chart. Which numbers are multiples of both 3 and 7?_____

14. Look at the 4s chart and the 9s chart. Which numbers are multiples of both 4 and 9?_____

15. Look at the 9s chart and the 6s chart. Which numbers are multiples of both 9 and 6?_____

Practice

Model Division

Complete the table. Use counters to help.

Counters	Number of Equal Groups	Number in Each Group
1. ●●●●●●●●● ●●●●●●●●●	_____	4
2. ●●●●●●●●●● ●●●●●●●●●● ●●●●●●●●●● ●●●●●●●●●●	8	_____
3. ●●●●●●● ●●●●●●● ●●●●●●● ●●●●●●●	_____	7
4. ●●●●●● ●●●●●● ●●●●●●	_____	6

Problem Solving and TAKS Prep

5. Gary has 45 stickers. He wants to put the same number of stickers on each of 9 pages. How many stickers will be on each page?

_____

6. Alice has 18 shells. She wants to put the same number of shells in each of 3 jars. How many shells will be in each jar?

_____

7. Which is the missing factor?

$7 \times \boxed{} = 21$

A 2
B 3
C 4
D 5

8. Al has 16 coins. He puts 4 coins in each of his boxes, using all 16 coins. How many boxes does Al have?

F 4
G 3
H 6
J 8

Practice

Relate Division and Subtraction

Write a division sentence for each.

1.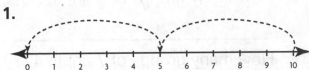

2.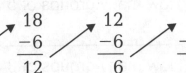

_____   _____

Use a number line or repeated subtraction to solve.

3. $12 \div 3 =$ _____

4. $20 \div 4 =$ _____

5. $21 \div 3 =$ _____

Problem Solving and TAKS Prep

6. Olivia went apple picking. She picked 48 apples. She put 6 apples in each of her baskets. How many baskets did Olivia use?

_____

7. Randy has 72 photographs. He puts his photographs into 8 equal piles. How many photographs are in each pile?

_____

8. Terri sets the table for 8 guests. She uses 16 plates. How many plates does each guest have?

   A 2
   B 24
   C 3
   D 8

9. Hal has 24 flowers in a bunch. He puts 4 flowers in each of his vases. How many vases does Hal use?

   F 8
   G 6
   H 20
   J 12

Practice

Model with Arrays

Use square tiles to make an array. Solve.

1. How many groups of 5 are in 25?   **2.** How many groups of 9 are in 36?

_____   _____

3. How many groups of 3 are in 12?   **4.** How many groups of 7 are in 42?

_____   _____

5. How many groups of 4 are in 16?   **6.** How many groups of 6 are in 24?

_____   _____

7. How many groups of 3 are in 18?   **8.** How many groups of 5 are in 35?

_____   _____

9. How many groups of 2 are in 14?   **10.** How many groups of 6 are in 54?

_____   _____

11. How many groups of 7 are in 21?   **12.** How many groups of 5 are in 40?

_____   _____

13. How many groups of 2 are in 18?   **14.** How many groups of 8 are in 16?

_____   _____

Make an array. Write a division sentence for each one.

15. 18 tiles in 6 groups   **16.** 28 tiles in 7 groups

_____   _____

17. George made an array with 70 tiles. He placed 7 tiles in each row.
How many rows did George make?_____

     Practice

© Harcourt

Algebra: Multiplication and Division

Complete.

1.

6 rows of ____ = 18

18 ÷ 6 = ____

2.

2 rows of ____ = 12

12 ÷ 2 = ____

3.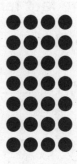

7 rows of ____ = 28

28 ÷ 7 = ____

Complete each number sentence. Draw an array to help.

4. $3 \times$ ____ $= 24$

   $24 \div 3 =$ ____

5. $4 \times$ ____ $= 32$

   $32 \div 4 =$ ____

6. $6 \times$ ____ $= 24$

   $24 \div 6 =$ ____

7. $9 \times$ ____ $= 36$

   $36 \div 9 =$ ____

Complete.

8. $3 \times 3 = 18 \div$ ____

9. $32 \div 8 =$ ____ $\times 2$

10. ____ $\times 1 = 35 \div 7$

Problem Solving and TAKS Prep

11. Karen has 15 tickets. A hot dog costs 5 tickets. What is the maximum number of hot dogs Karen can buy?

12. Molly is going to the movies with her friends. She has $40. Each ticket costs $8. What is the maximum number of tickets Molly can buy?

13. Tina has 30 baseball cards. She wants to divide them evenly between her 5 friends. How many cards will each friend receive?

    A 5   **C** 4

    B 6   **D** 7

14. The big fish tank has 42 fish. The fish will soon be divided evenly into 6 tanks. How many fish will be in each tank?

    F 5   **H** 7

    G 6   **J** 8

Practice

Algebra: Fact Families

Write the fact family for each set of numbers.

1. 4, 6, 24 _____ _____ _____ _____

2. 2, 9, 18 _____ _____ _____ _____

3. 5, 7, 35 _____ _____ _____ _____

Write the fact family for each array.

4. ●●●●●
   ●●●●●   _____   _____
   ●●●●●
   _____   _____

5. ●●●●●●●
   ●●●●●●●   _____   _____
   _____   _____

Problem Solving and TAKS Prep

6. Al buys a pack of watercolor paints that includes 12 colors. There are 2 colors in each of 6 rows. What is the fact family for the numbers 2, 6, and 12?

   _____ _____

   _____ _____

7. There are 18 cookies on a dish. There are 6 cookies in each of 3 rows on the dish. What is the fact family for the numbers 3, 6, and 18?

   _____ _____

   _____ _____

8. Which number sentence is NOT included in the same fact family as $7 \times 3 = 21$?

   A $21 \div 3 = 7$ **C** $21 \div 7 = 3$

   B $21 \times 3 = 7$ **D** $3 \times 7 = 21$

9. Which division sentence describes the array?

   F $2 \div 3 = 6$ **H** $3 \div 2 = 6$

   G $6 \div 2 = 3$ **J** $7 \div 7 = 1$

© Harcourt

Practice

Problem Solving Workshop Strategy: Write a Number Sentence

Problem Solving Strategy Practice

Solve. Write a number sentence for each.

1. Matt has 5 T-shirts. Adam has 12 T-shirts. How many more T-shirts does Adam have than Matt has?

_____

2. Isabelle has 8 books in her desk. She brought 4 more books from home and placed them in her desk. How many books does Isabelle have in her desk in all?

_____

3. A bag of marbles costs 60 cents. Each marble costs 10 cents. How many marbles are in the bag?

_____

4. There are 4 invitations in a box. Mrs. Hannah bought 8 boxes. How many invitations did Mrs. Hannah buy?

_____

Mixed Strategy Practice

5. **USE DATA** Tyler spent $40 on tickets. He bought 8 tickets of only one color. Which color were the tickets that Tyler bought?

Raffle Tickets	
Color	Cost
Yellow	$2
Green	$3
Blue	$5

_____

6. Mary spent $8 on a movie ticket, $12 on presents, and $15 on lunch. How much money did Mary spend in all?

7. Randall spent $75 of his, Marty's, and Jean's money on tickets. Randall used $25 of Marty's money and $12 of Jean's money. How much of his own money did Randall spend on tickets?

Practice

Divide by 2 and 5

Find each quotient.

1.

6 ÷ 2 = ____

2.

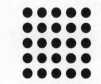

____ = 25 ÷ 5

3.

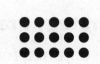

15 ÷ 5 = ____

4.

____ = 8 ÷ 2

5.

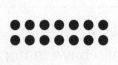

2)14

6.

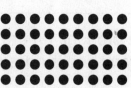

5)45

7.

2)2

8.

5)35

Problem Solving and TAKS Prep

9. Martin bought 40 packages of birdseed. He bought birdseed in 5-package cases. How many cases of birdseed did Martin buy?

_____

10. **Fast Fact** A female hummingbird usually lays 2 eggs. If a researcher finds 10 eggs in one area, how many female hummingbirds are most likely in the area?

_____

11. Sarah sees the same number of birds at each of 2 bird feeders. She sees 12 birds in all. How many birds does Sarah see at each bird feeder?

   A 4
   B 5
   C 6
   D 7

12. Greg has 5 bird feeders and a 20-pound bag of bird food. He puts the same amount of bird food into each feeder. How many pounds of bird food does Greg put into each feeder?

   F 3
   G 4
   H 5
   J 6

Name_____

Divide by 3 and 4

Find each quotient.

1.

2.

3.

4.

$12 \div 3 =$ ___   ___ $= 20 \div 4$ $21 \div 3 =$ ___   ___ $= 8 \div 4$

Complete.

5.

6.

7.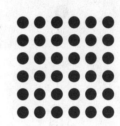

8.

$12 \div$ ___ $= 2$ $24 \div$ ___ $= 3$ $36 \div$ ___ $= 4$ $3 \div$ ___ $= 3$

Problem Solving and TAKS Prep

9. There are 24 students signed up for the relay race. Each team needs 4 students. How many teams will there be in the relay race?

_____

10. Twenty-one students form a study group. If they want to sit evenly at 3 tables, then how many students will be at each table?

_____

11. Jeremy has 36 crackers. He puts 4 crackers in each of his bags. How many bags does Jeremy have?

A 6
B 7
C 8
D 9

12. Lea has 27 beads. She makes 3 bracelets, each with the same number of beads. How many beads are on each bracelet?

F 9
G 8
H 7
J 6

© Harcourt

PW81

Practice

Division Rules for 1 and 0

Find each quotient.

1.

$5 \div 5 =$ _____

2.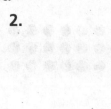

_____ $= 0 \div 4$

3.

$3 \div 1 =$ _____

4.

_____ $= 0 \div 9$

5.

$1\overline{)0}$

6.

$3\overline{)3}$

7.

$1\overline{)9}$

8.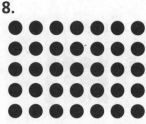

$5\overline{)35}$

Problem Solving and TAKS Prep

9. There are 7 stables at the Green Pastures Horse Farm. There are 7 horses that live on the farm. How many horses are in each stable, if there are an equal number of horses per stable?

_____

10. Trevor plans to give 3 grapes to each parrot in a store. There is 1 parrot in the store. How many grapes in all, does Trevor give to parrots in the store?

_____

11. Katherine has 5 birds. She only has 1 birdcage to keep them in. How many birds are in the cage?

A 0

B 1

C 5

D 10

12. Which is the quotient?

$4\overline{)0}$

F 0

G 1

H 2

J 4

Practice

Name_____

Algebra: Practice the Facts

Write a division sentence for each.

1.  2. 3.

_____   _____   _____

Find each missing factor and quotient.

4. $4 \times \boxed{} = 36$ $36 \div 4 =$ ___   5. $3 \times \boxed{} = 0$ $0 \div 3 =$ ___

Find each quotient.

6. $27 \div 3 =$ ___   7. $18 \div 3 =$ ___   8. $20 \div 4 =$ ___   9. ___ $= 32 \div 4$

10. $15 \div 5 =$ ___   11. $2 \div 2 =$ ___   12. $3 \overline{)21}$ 13. $2 \overline{)10}$

Problem Solving and TAKS Prep

14. A craft store sells beads in packages of 4. Tara needs 24 beads for a project. How many packages of beads will Tara need to buy?

15. Two brothers sell lemonade in their neighborhood. They make $6 on Saturday. How much money should each brother receive if they split this money evenly?

_____

16. Which division sentence is related to $3 \times 4 = 12$?

   A $24 \div 2 = 12$
   B $4 \div 2 = 2$
   C $12 \div 6 = 2$
   D $12 \div 3 = 4$

17. Which division sentence is related to $3 \times 8 = 24$?

   F $24 \div 3 = 8$
   G $24 \div 2 = 12$
   H $24 \div 6 = 4$
   J $24 \div 4 = 6$

Practice

Problem Solving Workshop Skill: Choose the Operation

Problem Solving Skill Practice

Choose the operation. Write a number sentence. Then solve.

1. The Murphy family spent $36 for 4 tickets to the nature center. How much did each ticket cost?

2. There were 27 children and 9 adults on the tour. How many people were on the tour in all?

3. The nature center has a petting zoo with 5 areas. Each area has the same number of animals. There are 25 animals in the petting zoo in all. How many animals are in each area?

4. Drinks at the nature center cost $7. Mr. Chin gave the clerk $20 for 1 drink. How much change should Mr. Chin get back from the clerk?

Mixed Applications

5. **USE DATA** Martha only hikes the Echo Trail. However she hikes this trail 3 times each week. How many miles does Martha hike each week?

Nature Trails	
Trail Name	**Distance**
Echo Trail	4 miles
View Trail	12 miles
Pine Trail	47 miles
Green Trail	15 miles
Steep Trail	23 miles

6. Cora, Sal, Marty, and Jane are standing in line. Jane is first in line. Marty is behind Cora. Cora is in front of Sal. Sal is behind Marty. In what order are the four people standing in line?

7. Anna needs 28 balloons. They come in packages of 4, 6, or 9 balloons. How many of each package should she buy in order to have the exact amount of balloons she needs?

© Harcourt

Practice

Divide by 6

Find each missing factor and quotient.

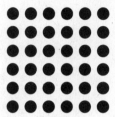

1. $6 \times \underline{\quad} = 42$ 2. $36 \div 6 = \underline{\quad}$ 3. $6 \times \underline{\quad} = 24$ 4. $\underline{\quad} \times 6 = 30$

Find each quotient.

5. $72 \div 6 = \underline{\quad}$

6. $24 \div 3 = \underline{\quad}$

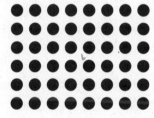

7. $\underline{\quad} = 48 \div 6$

8. $\underline{\quad} = 12 \div 6$

Problem Solving and TAKS Prep

9. Toni bought 24 hotdogs. They come in packages of 6. How many packages of hotdogs did Toni buy?

10. Kara brought 36 muffins to a picnic. Each package contains 6 muffins. How many packages of muffins did Kara bring?

_____

11. There are 42 books, divided evenly among six shelves in the bookcase. How many books are on each shelf?

   A 8   C 5
   B 6   D 7

12. There are 30 peaches in a basket. Frank separates the peaches evenly into 6 bags. How many peaches are in each bag?

   F 8   H 5
   G 6   J 7

Name_____

Divide by 7 and 8

Find each missing factor and quotient.

1. $8 \times \rule{1cm}{0.4pt} = 48$ **2.** $21 \div 7 = \rule{1cm}{0.4pt}$ **3.** $7 \times \rule{1cm}{0.4pt} = 28$ **4.** $\rule{1cm}{0.4pt} \times 8 = 40$

Find each quotient.

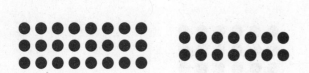
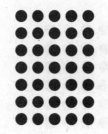

5. $24 \div 8 = \rule{1cm}{0.4pt}$ **6.** $14 \div 7 = \rule{1cm}{0.4pt}$ **7.** $\rule{1cm}{0.4pt} = 35 \div 7$ **8.** $\rule{1cm}{0.4pt} = 16 \div 2$

Problem Solving and TAKS Prep

9. The Williams family went camping at a lake. There are 56 members in the Williams family. Each cabin holds 8 people. How many cabins did the Williams family rent?

_____

10. Juana bought juice boxes for a camping trip. She needed 40 juice boxes. They come in packages of 8. How many packages of juice boxes did Juana buy?

_____

11. There were 56 apples in a cart. Don emptied the cart and put 7 apples into each of his bags. How many bags did Don fill?

A 12

B 7

C 8

D 6

12. Eva has 24 flowers. She arranges them into bunches of 8. How many bunches does Eva arrange?

F 6

G 24

H 8

J 3

Practice

Problem Solving Workshop Strategy: Work Backward

Problem Solving Strategy Practice

Work backward to solve.

1. Rachel spent $2.25 on a snack. Then her mom gave her $4.00. Now, Rachel has $9.25. How much money did Rachel have to start?

2. Abby cut a piece of construction paper into 2 equally long pieces. She then cut off 5 inches in length from one piece. This piece is now 4 inches long. What was the length of the original piece of construction paper?

_____

Mixed Strategy Practice

USE DATA For 3–4, use the table below.

3. Trent wants to buy a total of 10 shirts of either blue or green color. Are there enough shirts in inventory for him to buy the shirts he wants? Show your work.

T-Shirt Inventory	
Color	Number of Shirts
Blue	5
Green	4
Yellow	7
Red	6

_____

4. Frank bought 4 red T-shirts and 2 yellow T-shirts on Tuesday. If Frank wants to buy 16 more T-shirts on Wednesday, are there enough left in inventory for him to do so?

_____

5. Greg collected $81 selling 9 boxes of candy bars. How much did Greg charge for each box of candy bars?

6. Anna has 2 tambourines and 4 guitars. How many musical instruments does Anna have in all? Show your work.

_____

Divide by 9 and 10

Find each quotient.

1.

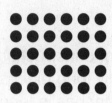

$30 \div 10 =$ _____

2.

$36 \div 9 =$ _____

3.

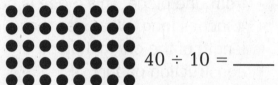

$40 \div 10 =$ _____

4.

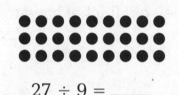

$27 \div 9 =$ _____

Complete each table.

5.

÷	40	60	80	100
10				

6.

÷	27	45	72	81
9				

Problem Solving and TAKS Prep

7. There are 54 fish, in 9 tanks, at an aquarium. Each tank contains an equal number of fish. How many fish are in each tank?

_____

8. A shark movie lasted for 50 minutes. The movie spent 10 minutes featuring each shark. How many sharks were featured in the movie?

_____

9. There are 40 people waiting in lines at an aquarium. There are 10 people in each line. How many lines are there?

   A 1

   B 4

   C 40

   D 400

10. Nine fish in a tank display a total of 36 stripes. If they each display an equal number of stripes, how many stripes does each fish display?

   F 9

   G 5

   H 6

   J 4

Divide by 11 and 12

Find each missing factor and quotient.

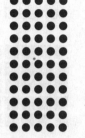

1. $11 \times \underline{} = 110$ **2.** $12 \times \underline{} = 108$ **3.** $55 \div 11 = \underline{}$ **4.** $96 \div 12 = \underline{}$

Find each quotient.

5. $11\overline{)132}$ **6.** $12\overline{)132}$ **7.** $10\overline{)120}$ **8.** $11\overline{)110}$

Problem Solving and TAKS Prep

9. Liam has 55 model cars. He places them evenly into 5 boxes. How many model cars are in each box?

_____

10. Allen bought 24 model train cars. There are 12 model train cars in a set. How many sets of model train cars did Allen buy?

_____

11. There are 72 tickets available for a show. If each person buys 12 tickets, how many people will it take to sell out the show?

A 6
B 7
C 8
D 9

12. David has 44 bottles; 11 bottles will fit on each shelf. How many shelves does David need?

F 2
G 3
H 4
J 5

Practice

Practice the Facts

Write a division sentence for each.

1.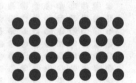

2.

3.
$$
\begin{array}{cccc}
48 & 36 & 24 & 12 \\
-12 & -12 & -12 & -12 \\
\hline
36 & 24 & 12 & 0
\end{array}
$$

_____   _____   _____

Find each missing factor and quotient.

4. $8 \times$ _____ $= 40$ $40 \div 8 =$ _____

5. $9 \times$ _____ $= 63$ $63 \div 9 =$ _____

Problem Solving and TAKS Prep

6. Thomas hiked on a trail that took 60 minutes to complete. Each section of the trail took 12 minutes to complete. How many sections does the trail have?

7. Carrie took 40 pictures on her nature walk. She took 4 pictures of every flower that she saw. How many flowers did Carrie see?

_____   _____

8. Hal walked 72 miles in 12 days. He walked the same number of miles each day. How many miles did Hal walk each day?

   A 3

   B 4

   C 5

   D 6

9. Nancy bought 4 new flashlights. Each flashlight cost $6.00. How much money did Nancy spend?

   F $18

   G $24

   H $30

   J $10

Practice

Line Segments and Angles

Tell whether each is a line, a line segment, or a ray.

1.

_____

2.

_____

3.

_____

4.

_____

5.

_____

6.

_____

7.

_____

8.

_____

Use the corner of a piece of paper to tell whether each angle
is *right*, *acute*, or *obtuse*.

9.

_____

10.

_____

11.

_____

12.

_____

Problem Solving and TAKS Prep

13. Bill wants to use toothpicks to
make a model of a stop sign.
How many line segments are in
a stop sign? Draw one here.

_____

14. Sally needs to be home at 3:00.
What type of angle is formed by
the two hands on a clock at 3:00?

_____

15. Which of the following appears to
be an obtuse angle?

16. Which of the following is a
line segment?

Practice

Types of Lines

Describe the lines. Tell if the lines appear to be *intersecting, perpendicular,* or *parallel*.

1.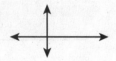

_____

2.

_____

3.

_____

4.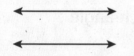

_____

5.

_____

6.

_____

Problem Solving and TAKS Prep

7. Marc wonders if every intersecting pair of lines is perpendicular. What would you tell him?

_____

_____

8. Can parallel lines be perpendicular lines as well? Why or why not?

_____

_____

9. Which of these pairs of lines appear to be parallel?

A

B

C

D

10. Which of these pairs of lines appear to be perpendicular?

F

G

H

J

Practice

Identify 2-Dimensional Figures

Name each figure. Tell how many sides.

1.

_____

2.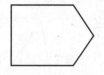

_____

3.

_____

4.

_____

5.

_____

6.

_____

7.

_____

8.

_____

For 9–11, use figures A–D.

9. Which of the figures have more than 3 sides?

_____

10. Which figure is a triangle?

_____

11. Which figure is a quadrilateral?

_____

Problem Solving and TAKS Prep

12. What plane figure has 6 sides and 6 vertices?

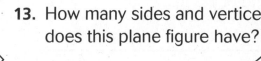

_____

13. How many sides and vertices does this plane figure have?

_____

14. How many sides does a quadrilateral have?

A 4   **C** 6

B 5   **D** 8

15. Which of the following plane figures is also a quadrilateral?

F ◯   **H**

G ◺   **J** ▢

© Harcourt

Triangles

Name each triangle. Write *equilateral, isosceles,* or *scalene.*

1.
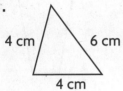
4 cm 6 cm
4 cm

_____

2.
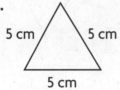
5 cm 5 cm
5 cm

_____

3.
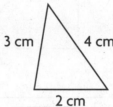
3 cm 4 cm
2 cm

_____

4.
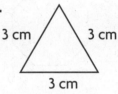
3 cm 3 cm
3 cm

_____

Name each triangle. Write *right, obtuse,* or *acute.*

5.
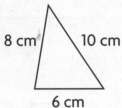
8 cm 10 cm
6 cm

_____

6.
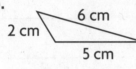
2 cm 6 cm
5 cm

_____

7.
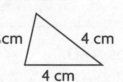
2 cm 4 cm
4 cm

_____

8.
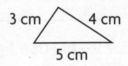
3 cm 4 cm
5 cm

_____

Problem Solving and TAKS Prep

9. A triangle has one side that is 3 cm long, one side that is 2 cm long, and one side that is 4 cm long. Two of the angles are acute and one angle is obtuse. What kind of triangle is it?

_____

10. Can a right triangle also be an isosceles triangle? Explain.

_____

_____

11. Which correctly names this triangle?

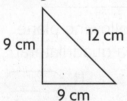
9 cm 12 cm
9 cm

A scalene, obtuse
B scalene, right
C isosceles, obtuse
D isosceles, right

12. Which correctly names this triangle?

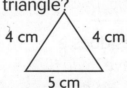
4 cm 4 cm
5 cm

F equilateral, acute
G scalene, obtuse
H isosceles, acute
J isosceles, obtuse

Practice

Quadrilaterals

Write as many names for each quadrilateral as you can.

1.

_____

_____

2.

_____

_____

3.

_____

_____

4.

_____

_____

5.

_____

_____

6.

_____

_____

Problem Solving and TAKS Prep

7. **Reasoning** A square is a rectangle. Is a rectangle a square? Explain.

_____

_____

_____

8. What type of quadrilateral has one pair of parallel sides but the sides are not always the same length?

_____

9. What type of quadrilateral is this figure?

A trapezoid
B rhombus
C square
D rectangle

10. Here is a quadrilateral. What two terms can be used to describe it?

F rectangle, parallelogram
G rhombus, parallelogram
H square, rectangle
J rhombus, square

Name_____

Circles

Name the gray part in each circle.

1.

_____

2.

_____

3.

_____

4.

_____

Is this gray part a radius? Write *yes* or *no*.

5.

_____

6.

_____

7.

_____

8.

_____

Problem Solving and TAKS Prep

9. Randy said, "a circle is a closed plane figure made of points that are the same distance from the radius." What single word can you replace in Randy's statement to make it true? What word should you replace it with?

_____

10. Dawn drew a circle and placed 2 gray points inside it. She said both points are centers. Can this be correct?

_____

11. Which of the following shows a radius in gray?

  A   C

  B   D

12. Which of the following shows only the center in gray and not a radius?

  F 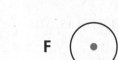   H

  G   J

Practice

Compare 2-Dimensional Figures

For 1–3, use the figures at the right.

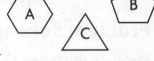

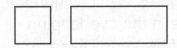

1. Which figures have only 3 sides? _____

2. Which figures have only 4 sides? _____

3. Which figures have parallel sides? _____

Problem Solving and TAKS Prep

4. How are an octagon and a triangle alike? How are they different?

5. How are a square and a rectangle alike? How are they different?

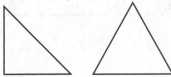

_____

_____

6. How are these figures alike?

7. How are these figures different?

A They both have 3 sides.

B They both have at least one right angle.

C They both have at least one obtuse angle.

D They both have 2 obtuse angles and one acute angle.

F They have a different number of sides.

G They both have one acute angle.

H Only one is a closed plane figure.

J They both have at least four obtuse angles.

Practice

Problem Solving Workshop Strategy:
Draw a Diagram

Problem Solving Strategy Practice

Draw a diagram to help solve the following problems.

1. Lance wanted to classify the following shapes into 2 categories: triangle, square, rhombus, rectangle, trapezoid, and octagon. He drew a Venn diagram and used the following headings for the 2 cirlces: "Parallelograms" and "Plane Figures." Which figures should be in the overlapping portion of his Venn Diagram?

_____

2. Nineteen students take music classes. Three students take only the trumpet class. Six students take only the piano class. Ten students take both the trumpet and the piano class. How many students take the piano class in all?

_____

Mixed Strategy Practice

3. **USE DATA** How many crew members are on 6 boats? Look for a pattern to solve.

Boats	2	3	4	5	6
Crew Members	8	12	16	20	?

_____

4. Lauren ate apples three days in a row. On Monday she ate 6 apples. On Tuesday she ate 3 apples. On Wednesday she ate 1 apple. How many apples did Lauren eat in all? Show your work.

_____

Practice

Congruent Figures

Tell if the figures appear to be congruent. Write *yes* or *no*.

1.   2.   3.

_____   _____   _____

4.   5.   6.

_____   _____   _____

For 7–8, use the figures in the chart.

7. Katie drew a model of her school. Which figure appears to be congruent to the model Katie drew?

_____

8. Michael goes to a different school. He also drew a model. Which figure appears to be congruent to Michael's model?

_____

Katie's model

Michaels's model

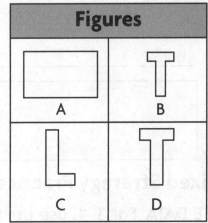

Figures	
A	B
C	D

Problem Solving and TAKS Prep

9. Jason drew the two figures below. Do the two figures appear to be congruent?

_____

10. Mary drew the two figures below on a sheet of paper. Do the two figures appear to be congruent?

_____

11. Which figure appears to be congruent to the gray figure?

A   C

B   D

12. Which item appears to show two congruent figures?

F   H

G   J

Name_____

Problem Solving Workshop Strategy: Make a Model

Problem Solving Strategy Practice

Make a model to solve.

For 1–2, use the pattern blocks at the right.

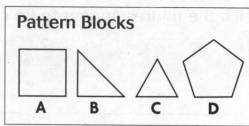
Pattern Blocks

A   B   C   D

1. Karen used 2 pattern blocks to make a shape that appears to be congruent to the one below. Which pattern blocks did she use?

2. John used 2 pattern blocks to make a parallelogram. Which pattern blocks did he use?

Mixed Strategy Practice

USE DATA For 3–4, use the table.

3. How many square-shaped and triangle-shaped tiles are there in all? Show your work.

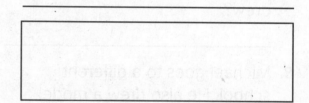

Mosaic Tile Kit	
Tile Shape	**Number in Kit**
Square	80
Rectangle	74
Triangle	55
Trapezoid	32

4. The squares are either red or green. There is an equal number of each color. How many squares are red?

5. Kyle made an array using 15 square tiles. His array had 3 columns. How many rows were in Kyle's array?

Practice

© Harcourt

Symmetry

Tell if the gray line appears to be a line of symmetry.
Write *yes* or *no*.

1.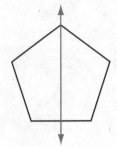

2.

3.

4.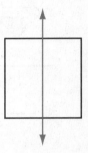

_____   _____   _____   _____

Problem Solving and TAKS Prep

5. **Reasoning** Andrew wants to cut an apple in half. Explain how he can use the line of symmetry to do this.

_____

_____

_____

6. **Reasoning** Does the figure at the right appear to have a line of symmetry? Explain.

_____

_____

_____

7. Which appears to show a line of symmetry?

A

C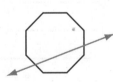

B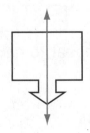

D

8. Which does NOT appear to show a line of symmetry?

F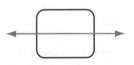

H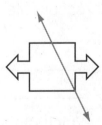

G

J

Practice

Lines of Symmetry

Draw the line or lines of symmetry for each figure.

1.

2.

3.

4.

5.

6.

7.

8.

Decide if each figure appears to have 0 lines, 1 line, or more than 1 line of symmetry. Write *0, 1,* or *more than 1*.

9.

_____

10.

_____

11.

_____

12.

_____

Problem Solving and TAKS Prep

13. **Reasoning**
Nancy went to the beach and found the sea star at the right. She decided that the sea star does not have a line of symmetry. Is Nancy's decision reasonable? Explain.

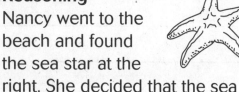

_____

_____

_____

14. How many lines of symmetry does the figure at the right appear to have? Explain.

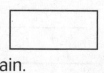

_____

_____

_____

15. Which of the following letters appears to have more than one line of symmetry?

A **R**   c **A**

B **K**   D **X**

16. How many lines of symmetry does the figure on the right appear to have?

F 0   H 2

G 1   J 3

Practice

Draw Symmetric Figures

Complete the design so each has a line of symmetry.

1. 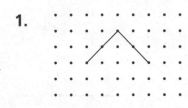 2. 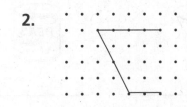 3.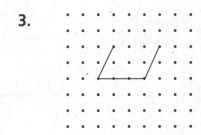

For 4–5, use the equilateral triangles below.

4. How many lines of symmetry does an equilateral triangle have? _____

5. How many lines of symmetry does the
   double triangle appear to have?

 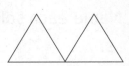

_____

Problem Solving and TAKS Prep

6. Taryn drew the lines of
   symmetry on the octagon below.
   Did she draw them all? If not,
   draw the one(s) you think
   she missed.

   _____

7. Does this figure appear to have a
   line of symmetry? If so, draw it.

   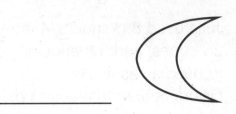

   _____

8. Which figure appears to show a
   line of symmetry?

   A   C

   B   D

9. Which figure appears to show a
   line of symmetry?

   F   H

   G   J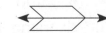

Practice

Name_____

Identify 3-Dimensional Figures

Name the solid figure that each object is shaped like.

1.

2.

3.

4.

_____ _____ _____ _____

5.

6.

7.

8.

_____ _____ _____ _____

Name each solid figure.

9.

10.

11.

12.

_____ _____ _____ _____

Problem Solving and TAKS Prep

13. Julia used 8 cylinders, 4 cones, 25 cubes, and 3 triangular prisms to build a castle. How many more cylinders than triangular prisms did Julia use?

_____

14. Roger used 4 cylinders, 3 cones, 12 cubes, and 1 sphere to build a tower. Half the figures were blue and half were red. How many figures were red?

_____

15. Which solid figure is the tent below shaped like?

A cone
B cube
C triangular prism
D rectangular prism

16. Which solid figure is the book below shaped like?

F cube
G square pyramid
H rectangular prism
J triangular prism

Practice

Faces, Edges, and Vertices

Name the solid figure. Then tell how many faces, edges, and vertices.

1.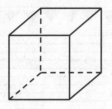

_____

_____ faces

_____ edges

_____ vertices

2.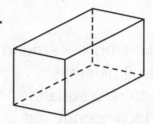

_____

_____ faces

_____ edges

_____ vertices

3.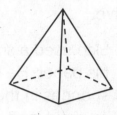

_____

_____ faces

_____ edges

_____ vertices

Name the solid figure that has the faces shown.

4.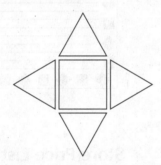

_____

5.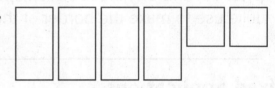

_____

Problem Solving and TAKS Prep

6. Rene made the birdfeeder at the right from a plastic box. How many faces and how many vertices does the birdfeeder have?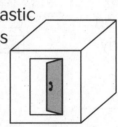

_____

7. Gwynn makes a wooden model of a tent. The tent is in the shape of a square pyramid. How many faces does Gwynn's model have?

_____

8. Which solid figure is shaped like a drinking straw?

   A cone   **C** cylinder

   B cube   **D** sphere

9. Which represents the number of edges that a small cube has?

   F 8   **H** 4

   G 6   **J** 12

Practice

Problem Solving Workshop Skill: Identify Relationships

Problem Solving Skill Practice

Solve.

1. Skip used a sponge to make a border around his paper. He had access to 3 different sponges in the shape of a cube, a square pyramid, and a cylinder. Which sponge did Skip use to make the border at the right?

_____

2. Julie used sponges to make a border around her paper. She had access to 3 different sponges in the shape of a cube, a square pyramid, and a cylinder. Which sponges did Julie use to make the border at the right?

_____

Mixed Applications

USE DATA For 3–4, use the Store Price List below.

3. Alice is told to spend exactly $13 at the store on two items. What will the shape of the two items be?

_____

Store Price List

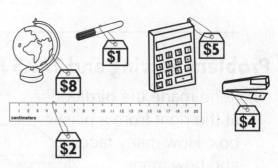

4. Cindy went to the movies on Saturday and spent $8. She went to the store on Monday and spent $8. What did Cindy buy at the store?

_____

5. Bobby collected 8 baseball cards and 2 basketball cards. He put the cards evenly in each of 5 cylinder shaped canisters. How many cards were in each canister? Show your work.

_____

_____

Practice

Compare 3-Dimensional Figures

Compare the figures. Tell one way they are alike or different.

1.

_____

2.

_____

3.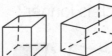

_____

For 4–6, identify each figure.

4. I have 5 faces. Four of my faces are triangles. Which solid figure am I?

_____

5. Only two of my faces are triangles, the rest are of another shape. Which solid figure am I?

_____

6. I have 6 faces. All my faces are square shaped. Which solid figure am I?

_____

7. I am a solid figure with a curved surface. Which solid figure am I?

_____

Problem Solving and TAKS Prep

8. Pedro and June make clay models. Pedro's model is shaped like a square pyramid. June's model is shaped like a cube. How are the two figures alike?

_____

9. Mia and Sue make paper models. Mia's model is shaped like a cube. Sue's model is shaped like a rectangular prism. How are the two figures alike?

_____

10. How are a triangular prism and a square pyramid alike?

   A Both have 5 faces.

   B Both have 8 edges.

   C Both have 6 vertices.

   D Both have no faces.

11. I have 5 faces. Which solid figure am I?

   F cylinder

   G square pyramid

   H sphere

   J cone

Practice

Compare Attributes

Compare.

1. Which is longer?

_____

2. Which is shorter?

_____

3. Which is lighter?

_____

4. Which holds more?

_____

5. Which is heavier?

_____

6. Which holds less?

_____

7. Draw an object that is heavier than a book. Explain your choice.

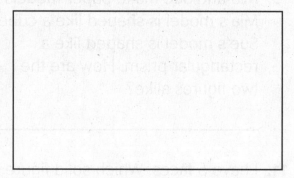

_____

8. Draw an object that is shorter than you. Explain your choice.

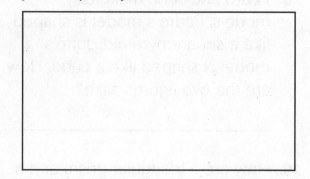

_____

Practice

Name_____

Length

Choose the unit you would use to measure each. Write *inch, foot, yard* or *mile*.

1.

2.

3.

_____ _____ _____

4. the length of a cereal box

5. the length of a spoon

6. the length of the Mississippi River

_____ _____ _____

7. the length of a tea kettle

8. the distance between the north side of two cities

9. the length of an automobile

_____ _____ _____

Problem Solving and TAKS Prep

10. Justin plans to hike for many hours through the mountains. Which customary unit of length best describes how far Justin will hike?

11. Alex saw an adult shark at the aquarium. Which customary unit of length best describes the length of the shark?

_____ _____

12. Lilly wants to measure the length of a bike. About how long is the bike?

   A 5 inches

   B 5 feet

   C 5 yards

   D 5 miles

13. Tyler wants to measure the length of a book. About how long is the book?

   F 9 inches

   G 9 feet

   H 9 yards

   J 9 miles

Practice

Name_____

Estimate and Measure Inches

Measure the length to the nearest inch.

1.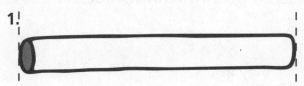

_____

2.

_____

Measure the length to the nearest half inch.

3.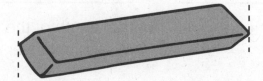

_____

4.

_____

Use a ruler. Draw a line for each length.

5. $2\frac{1}{2}$ inches

6. 1 inch

Problem Solving and TAKS Prep

7. Nina measures a marker that is $2\frac{1}{2}$ inches long. Between which two inch-marks does the end of the marker lie?

_____

8. What is the length of the card below to the nearest half inch?

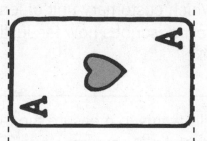

_____

9. Which is the length of the string below to the nearest half inch?

A 1 inch   **C** 2 inches

B $1\frac{1}{2}$ inches   **D** $2\frac{1}{2}$ inches

10. Which is the length of the string below to the nearest half-inch?

F 2 inches   **H** 3 inches

G $2\frac{1}{2}$ inches   **J** $3\frac{1}{2}$ inches

© Harcourt

PW110

Practice

Estimate and Measure Feet and Yards

Choose the better unit of measure.

1. the length of a rug

   8 feet or 8 inches

2. the length of a puppy

   1 foot or 1 yard

3. the length of a soccer field

   100 feet or 100 yards

4. the length of a pickup truck

   5 feet or 5 yards

5. the length of a sofa

   6 feet or 6 yards

6. the length of a tennis court

   80 feet or 80 yards

Use the Table of Measures. Write the length in feet and inches or in yards and feet.

7. 38 inches = ☐ feet ☐ inches

8. 14 feet = ☐ yards ☐ feet

9. 42 inches = ☐ yard ☐ inches

10. 102 inches = ☐ feet ☐ inches

11. 8 feet = ☐ yards ☐ feet

Table of Measures
1 foot = 12 inches
1 yard = 3 feet
1 yard = 36 inches

12. Jamie plans to knit a sweater. She needs 12 feet of yarn. She has 3 yards of yarn. Does Jamie have enough yarn to knit the sweater? Explain.

_____

_____

_____

Capacity

Choose the unit you would use to measure each. Write *cup, pint, quart,* or *gallon*.

1.

2.

3.

4.

_____ _____ _____ _____

5.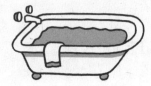

6.

7.

8.

_____ _____ _____ _____

Tell how the units are related.

9. 10 cups = ☐ pints

10. 4 quarts = ☐ pints

11. ☐ quarts = 2 gallons

12. 4 gallons = ☐ cups

13. ☐ pints = 7 quarts

14. 12 cups = ☐ quarts

15. 14 pints = ☐ cups

16. ☐ quarts = 1 gallon

17. 10 gallons = ☐ quarts

18. Beth needs to bring a gallon of juice to a party. She bought 2 quarts of juice. She has no other juice in her possession other than the juice she just bought. Did Beth buy enough juice? Explain.

_____

_____

Practice

Weight

Choose the unit you would use to weigh each. Write *ounce* or *pound*.

1.

_____

2.

_____

3.

_____

4.

_____

5.

_____

6.

_____

7.

_____

8.

_____

Find two objects in the classroom to match each weight.
Draw them and label their weights.

9. about 5 pounds

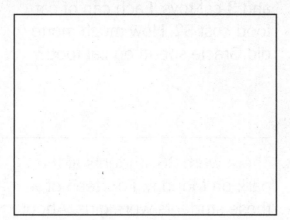

10. about 4 ounces

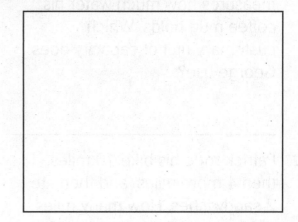

11. Sam told his friend that his two month old puppy weighs 4. He did not give the unit. Which unit of weight should Sam have said after 4, ounces or pounds?

_____

Problem Solving Workshop Skill: Choose a Unit

Problem Solving Skill Practice

Solve by choosing the better unit of measure.

1. Mr. Brill wants to measure the distance from each goal line to the half-field line of a soccer field. Which customary unit of length should Mr. Brill use?

2. Allison makes juice for herself and her 3 friends. Which customary unit of capacity does Allison use to measure the amount of juice she makes?

3. George measures how much water his kitchen sink holds. Which customary unit of capacity does George use?

4. Julie measures the length of her sister's hair. Which customary unit of length does Julie use?

Mixed Applications

5. **Pose a Problem** George measures how much water his coffee mug holds. Which customary unit of capacity does George use?

6. Gracie bought 6 cans of cat food, and 3 cat toys. Each can of cat food cost $2. How much money did Gracie spend on cat food?

7. Patrick rode his bike 10 miles, then 4 more miles, and then ate 2 sandwiches. How many miles did Patrick ride his bike in all?

8. There were 26 students at the park on Monday. Fourteen of these students were girls. About how many students, at the park on Monday were boys?

Practice

Name_____

Fahrenheit Temperature

Write each temperature in °F.

1.

2.

3.
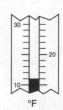

4.

_____   _____   _____   _____

Choose the better temperature for each activity.

5.

6.

7.

8.

28°F or 78°F   82°F or 32°F   65°F or 25°F   53°F or 93°F

_____   _____   _____   _____

9. It is 27°F outside. What is an activity Jeanne might be doing outside? What clothes do you think Jeanne might wear for this activity?

_____

_____

_____

Practice

Name_____

Use a Thermometer

Use the thermometers. Find the difference in temperatures.

1.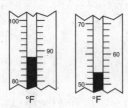

_____

2.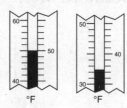

_____

3.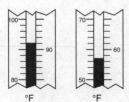

_____

4.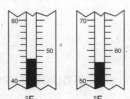

_____

Problem Solving and TAKS Prep

For 5–6, use the thermometers.

5. How much did the temperature rise from 6 A.M. to noon?

_____

6. The temperature at midnight was 12°F cooler than it was at noon. What was the temperature at midnight?

_____

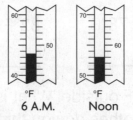

°F   °F
6 A.M.   Noon

7. The temperature at 12:00 P.M. was 67°F, which was 9°F warmer than at 9:00 P.M. What was the temperature at 9:00 P.M.?

   A 57°F   C 67°F

   B 58°F   D 76°F

8. The temperature at 6:00 A.M. was 34°F, which was 11°F cooler than at 1:00 P.M. What was the temperature at 1:00 P.M.?

   F 23°F   H 44°F

   G 25°F   J 45°F

Practice

Length

Choose the unit you would use to measure each.
Write *cm, m,* or *km.*

1.

_____

2.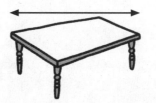

_____

3.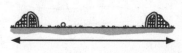

_____

4.

_____

5.

_____

6.

_____

7. distance between the north side of two towns

_____

8. width of a book

_____

9. height of a building

_____

10. length of a fire truck

_____

11. distance to the moon

_____

12. length of your hand

_____

Problem Solving and TAKS Prep

13. Sheila wants to measure the distance between first and second base on a baseball field. Which unit should Sheila use?

_____

14. Pedro hit a home run. Did the ball travel 90 cm, 90 dm, 90 m, or 90 km?

_____

15. Which has a length of about 1 decimeter?

    A a hockey stick

    B a crayon

    C a paper clip

    D your big toe

16. Which unit would you use to measure the length of your classroom?

    F cm

    G dm

    H m

    J km

Practice

Centimeters and Decimeters

Estimate the length in centimeters. Then use a centimeter ruler to measure to the nearest centimeter.

1.

_____

2.

_____

3.

_____

4.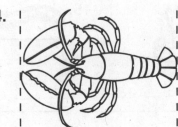

_____

Circle the better estimate.

5.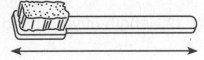

16 cm or 16 dm

6.

13 cm or 13 m

Problem Solving and TAKS Prep

7. Leo is 10 dm tall. Lauren is 98 cm tall. Who is taller?

_____

8. A tree in Miguel's front yard is 80 dm tall. How many centimeters tall is the tree?

_____

9. Shirley measured the length of her math book. Which could be the length of the book?

   A 600 cm   **C** 26 cm

   B 16 dm   **D** 46 dm

10. Which object is about 1 dm tall?

   F giraffe   **H** light post

   G cow   **J** soup can

Practice

Name_____

Meters and Kilometers

Choose the unit you would use to measure each. Write *m* or *km*.

1.

_____

2.

_____

3.

_____

4.

_____

5.

_____

6.

_____

7.

_____

8.

_____

9.

_____

Problem Solving and TAKS Prep

10. The world's tallest mountain is Mount Everest in the Himalayas in Asia. It is about 8,848 meters tall. Is Mount Everest taller or shorter than 9 kilometers? By how many meters?

_____

11. The tallest mountain in North America is Mount McKinley in Alaska. It is about 96 meters taller than 6 kilometers. About how many meters tall is Mount McKinley?

_____

12. If Mr. Smith takes 4 hours to drive to Benton from home, and he drives 100 km per hour. About how many kilometers away from Mr. Smith's home is Benton?

A 4   **C** 400

B 40   **D** 4,000

13. Which is about 1 meter long?

F book   **H** river

G pencil   **J** umbrella

Practice

Name_____

Lesson 20.4

Capacity

Choose the unit you would use to measure the capacity of each.
Write *mL* or *L*.

1.

2.

3.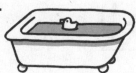

4.

_____ _____ _____ _____

5.

6.

7.

8.

_____ _____ _____ _____

9.

10.

11.

12.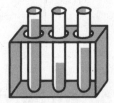

_____ _____ _____ _____

13. In the space at the right, draw and label a picture of a container that has a capacity less than 1 liter.

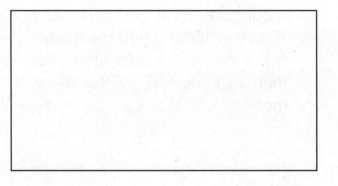

Find each missing number.

14. _____ mL = 3 L

15. _____ L = 6,000 mL

16. 9,000 mL = _____ L

17. 10 L = _____ mL

18. 20,000 mL = _____ L

19. _____ L = 13,000 mL

PW120

Name_____

Mass

Choose the unit you would use to find the mass of each. Write *gram* or *kilogram*.

1.

_____

2.

_____

3.

_____

4.

_____

5.

_____

6.

_____

7.

_____

8.

_____

9.

_____

10.

_____

11.

_____

12.

_____

13. In the space at the right, draw and label an object that has a mass greater than 1 kilogram.

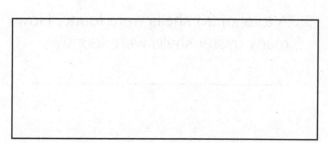

Find each missing number.

14. _____ g = 6 kg

15. 12,000 g = _____ kg

16. 20 kg = _____ g

Practice

Problem Solving Workshop Strategy: Compare Strategies

Problem Solving Strategy Practice

Make a table or act it out to solve.

1. Belinda found a horseshoe crab on the beach. The horseshoe crab measured 40 cm from the tip of its tail to the top of its head. How many decimeters long was the horseshoe crab?

2. Silas made a sand castle with a ditch around it. He poured 3 L of seawater into the ditch. How many milliliters of seawater did Silas pour into the ditch?

Mixed Strategy Practice

3. Lucia can carry 4,000 mL of seawater in her pail. How many liters of seawater can Lucia carry in her pail?

_____

4. Together, Belinda and Silas collected 40 seashells. Belinda collected 10 more seashells than Silas. How many seashells did each person collect? Guess and check to solve.

USE DATA For 5–6, use the graph.

5. A total of 23 shells were found. How many oyster shells were found?

_____

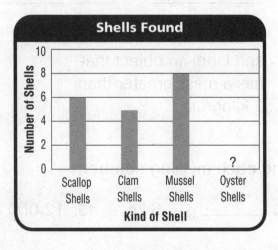

6. Lucia wanted to save two kinds of shells listed in the graph. What combinations of shells could she choose to save?

_____

_____

Practice

Perimeter

Find the perimeter of each figure.

1.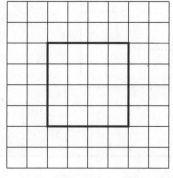

_____

2.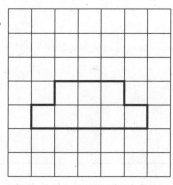

_____

3.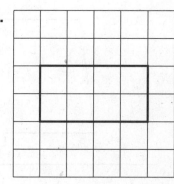

_____

4.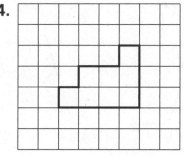

_____

5.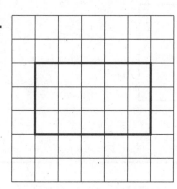

_____

6.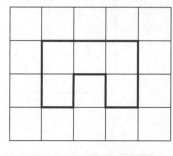

_____

7.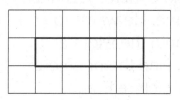

_____

8.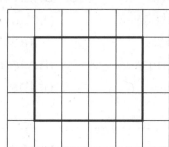

_____

9.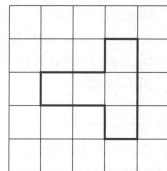

_____

Estimate and Measure Perimeter

Estimate. Then use a centimeter ruler to find the perimeter.

1.

_____

2.

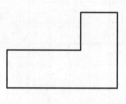

_____

Estimate. Then use an inch ruler to find the perimeter.

3.

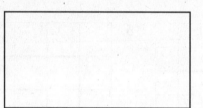

_____

4.

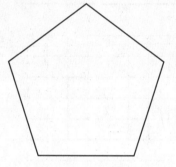

_____

Problem Solving and TAKS Prep

5. John has a 5-inch by 7-inch picture frame and a 4-inch by 6-inch picture frame. Which picture frame has the greater perimeter?

_____

6. Brian has an 8-inch by 10-inch picture frame. He wants to add 1 inch to both the width and the length. Find the perimeter of the new picture frame.

_____

7. What is the perimeter of this triangle?

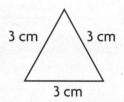

A 6 cm   **C** 9 cm

B 8 cm   **D** 12 cm

8. The figure below has a perimeter of 20 cm. How long is the fourth side?

F 2 cm   **H** 10 cm

G 8 cm   **J** 12 cm

Practice

Area of Plane Figures

Find the area of each figure. Write the answer in square units.

1.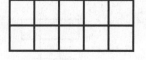

2.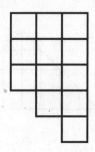

3.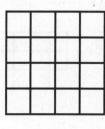

_____   _____   _____

4.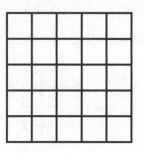

5.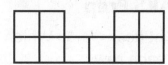

6.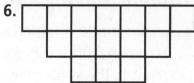

_____   _____   _____

7.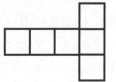

8.

9.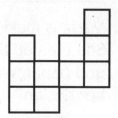

_____   _____   _____

Practice

Name_____

Find Area

Count or multiply to find the area of each figure. Write the answer in square units.

1.

2.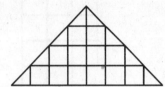

3.

_____   _____   _____

Problem Solving and TAKS Prep

4. Leslie made a tabletop that has 10 rows with 8 blocks in each row. What is the area of the tabletop in square units?

5. Which figure has the greater area?

Figure A   Figure B

_____

_____

6. Maria is making a table runner. She created the design below on grid paper. What is the area of her design?

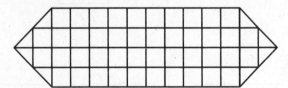

   A 28 square units

   B 32 square units

   C 48 square units

   D 52 square units

7. Paul is making a hot plate with red and white tiles. He has 6 rows of tiles with 6 tiles in each row. What is the area of Paul's hot plate?

   F 18 square units

   G 20 square units

   H 34 square units

   J 36 square units

Practice

Relate Perimeter and Area

For each pair, find the perimeter and the area.
Tell which figure has the greater area.

1.

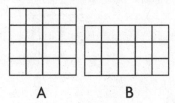

2.

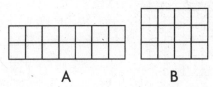

_____   _____

_____   _____

_____   _____

Problem Solving and TAKS Prep

3. Leah is making a picture frame. The perimeter of her picture is 24 inches and the area is 35 square inches. What are the lengths of the picture's sides?

4. Luke's garden has a perimeter of 16 feet. Which design will give his garden the greatest area?

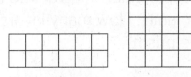

   Design A   Design B

_____   _____

5. Which figure has an area of 12 square units?

A    C

B   D

6. Which figure has a perimeter of 14 units?

F 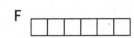   H

G 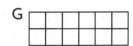   J

Practice

Volume

Use cubes to make each solid. Then write the volume in cubic units.

1.

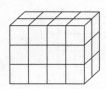

_____

2.

_____

3.

_____

4.

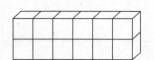

_____

5.

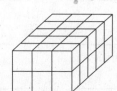

_____

6.

_____

Problem Solving and TAKS Prep

7. Each layer of a rectangular prism is 4 cubic units. The volume is 8 cubic units. How many layers are in the prism?

_____

8. Teresa has 18 cubes to make a solid figure with 6 cubes in each layer. How many layers will the solid figure have?

_____

9. What is the volume of this solid figure below?

A 12 cubic units

B 18 cubic units

C 27 cubic units

D 30 cubic units

10. What is the volume of the solid figure below?

F 3 cubic units

G 6 cubic units

H 9 cubic units

J 12 cubic units

Practice

Problem Solving Workshop Skill: Use a model

Problem Solving Skill Practice

Use a model to solve.

USE DATA For 1–3, use the boxes at the right.

1. Lillian keeps her ornaments in cubed-shaped boxes. She has 2 large boxes of ornaments. She is looking for a special ornament that is in a box that holds 40 ornaments. In which box should she look?

   _____

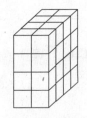 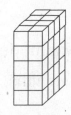

Box A   Box B

2. What if Box B could hold only 1 layer of cube-shaped ornament boxes? What would be the volume of Box B in cubic units?

   _____

3. What if Box A could only hold 3 layers of cube-shaped ornament boxes? What would be the volume of Box A in cubic units?

   _____

Mixed Applications

4. Tom has two cartons of golf balls. Carton A has 3 layers with 15 golf balls in each layer. Carton B has 4 layers with 12 golf balls in each layer. Which carton holds the greater amount of golf balls?

   _____

5. Ella is buying a case of pears. Each row has 10 pears and there are 3 rows. If the cost of a pear is $0.50, how much will the case of pears cost in all?

   _____

6. Wesley has 4 more hockey cards than baseball cards. If he has 28 cards in all, how many hockey cards does he have?

   _____

7. I am a 2-digit number. My tens digit is two more than my ones digit. My ones digit is between 4 and 6. What number am I?

   _____

Practice

Model Part of a Whole

Write a fraction in numbers and in words that names the shaded part.

1.

2.

3.

_____   _____   _____

Draw a model of each. Then write the fraction using numbers.

4. two fifths

5. seven tenths

6. five out of eight

_____   _____   _____

Problem Solving and Test Prep

7. Sam cut the apple pie into 6 pieces. He ate one slice of pie. What fraction of the pie is left?

_____

8. Sam gave Jenny 2 slices of pie. Now, what fraction of the pie is left?

_____

9. What fraction of the figure has been shaded?

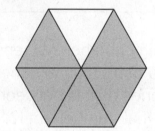

_____

10. What fraction of the figure has been shaded?

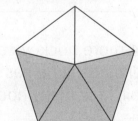

A $\frac{2}{5}$ C $\frac{1}{3}$

B $\frac{3}{3}$ D $\frac{3}{5}$

Practice

Model Part of a Group

Write a fraction that names the part of each group that is black.

1.

_____

2.

_____

3.

_____

4.

_____

Draw each. Then write the fraction that names the shaded part.

5. Draw 5 squares.
   Shade 2 squares.

6. Draw 8 circles.
   Shade 5 circles.

7. Draw 4 diamonds.
   Shade 3 diamonds.

_____

_____

Problem Solving and TAKS Prep

USE DATA For 8–9, use the bar graph.

8. The bar graph shows the marbles in Addy's collection. How many marbles does she have? _____

9. What fraction of the marbles are brown?

_____

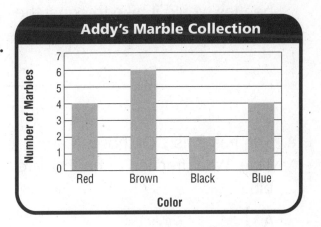

10. What fraction of the coins are dimes?

A $\frac{1}{2}$

B $\frac{1}{8}$

C $\frac{4}{8}$

D $\frac{2}{8}$

11. Jack has 10 toy trucks. $\frac{1}{5}$ of Jack's toy trucks are red. How many trucks are red?

F 2

G 5

H 1

J 10

Equivalent Fractions

Find an equivalent fraction. Use fraction bars.

1.

$\frac{1}{3}$	$\frac{1}{3}$

2.

$\frac{1}{6}$	$\frac{1}{6}$	$\frac{1}{6}$	$\frac{1}{6}$

3.

$\frac{1}{5}$	$\frac{1}{5}$	$\frac{1}{5}$

_____   _____   _____

Find the missing numerator. Use fraction bars.

4. $\frac{2}{8} = \frac{\square}{4}$

5. $\frac{1}{2} = \frac{\square}{10}$

6. $\frac{3}{3} = \frac{\square}{6}$

Problem Solving and TAKS Prep

7. USE DATA The bar graph shows the fictional weights of three different types of bugs. How many beetles would it take to equal the weight of one dragonfly?

Bugs	
Type	**Weight**
Beetle	$\frac{1}{8}$ gram
Grasshopper	$\frac{1}{2}$ gram
Dragonfly	$\frac{3}{4}$ gram

_____

8. Erin has 8 fish. Of these, 3 are blue. What fraction of her fish are blue?

  A $\frac{1}{2}$

  B $\frac{5}{8}$

  C $\frac{3}{8}$

  D $\frac{2}{4}$

9. What is the missing numerator?

$$\frac{4}{12} = \frac{\blacksquare}{3}$$

  F 2

  G 4

  H 1

  J 6

Practice

Simplest Form

Write each fraction in simplest form. Use fraction bars or counters.

1.

| $\frac{1}{8}$ | $\frac{1}{8}$ | $\frac{1}{8}$ | $\frac{1}{8}$ | $\frac{1}{8}$ | $\frac{1}{8}$ |

| $\frac{1}{4}$ | $\frac{1}{4}$ | $\frac{1}{4}$ |

2.

| $\frac{1}{6}$ | $\frac{1}{6}$ | $\frac{1}{6}$ | $\frac{1}{6}$ |

| $\frac{1}{3}$ | $\frac{1}{3}$ |

3.

| $\frac{1}{10}$ | $\frac{1}{10}$ | $\frac{1}{10}$ | $\frac{1}{10}$ | $\frac{1}{10}$ |

| $\frac{1}{2}$ |

$$\frac{6}{8} = \frac{\square}{\square}$$

$$\frac{4}{6} = \frac{\square}{\square}$$

$$\frac{5}{10} = \frac{\square}{\square}$$

4. $\frac{2}{8} = \frac{\square}{\square}$ **5.** $\frac{6}{9} = \frac{\square}{\square}$ **6.** $\frac{4}{12} = \frac{\square}{\square}$ **7.** $\frac{10}{14} = \frac{\square}{\square}$ **8.** $\frac{3}{9} = \frac{\square}{\square}$ **9.** $\frac{9}{12} = \frac{\square}{\square}$

Problem Solving and TAKS Prep

10. Trevor ate 2 out of 8 pieces of candy. What fraction of the candy is left? Write your answer in simplest form.

11. Jen read 8 out of 12 pages of the chapter. What fraction of the pages did she read? Write your answer in simplest form.

_____

12. What fraction is in simplest form?

A $\frac{1}{2}$

B $\frac{4}{8}$

C $\frac{3}{12}$

D $\frac{2}{6}$

13. What is $\frac{3}{9}$ in simplest form?

F $\frac{2}{3}$

G $\frac{1}{6}$

H $\frac{1}{3}$

J $\frac{1}{9}$

Practice

Fractions on a Number Line

Write the missing fraction(s) for each number line.

1.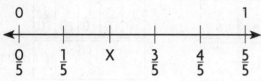

2.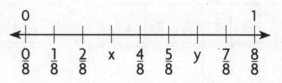

_____   _____

Tell which point represents each fraction.

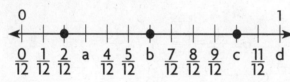

3. $\frac{3}{12}$

4. $\frac{12}{12}$

5. $\frac{1}{2}$

6. $\frac{5}{6}$

_____ _____ _____ _____

Problem Solving and TAKS Prep

USE DATA For 7–8, use the number line below.

7. The number line shows how far (in miles) Kate, Ryan, and Amy live from school. How far does Kate live from school?

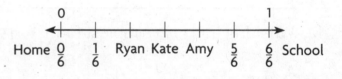

_____

8. How much farther does Amy live from school than Ryan?

_____

9. Andy has 12 coins. He wants to put them into 3 equal groups. How many coins will be in each group?

   A 1   **C** 3

   B 4   **D** 2

10. Which point represents $\frac{2}{5}$ on the number line?

   F a

   G c

   H b

   J d

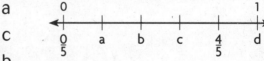

Practice

© Harcourt

Compare Fractions

Compare. Write <, >, or = for each ◯.

1.

| $\frac{1}{8}$ | $\frac{1}{8}$ | $\frac{1}{8}$ | $\frac{1}{8}$ |

| $\frac{1}{3}$ | $\frac{1}{3}$ |

$\frac{4}{8}$ ◯ $\frac{2}{3}$

2.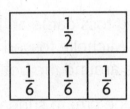

| $\frac{1}{2}$ |

| $\frac{1}{6}$ | $\frac{1}{6}$ | $\frac{1}{6}$ |

$\frac{1}{2}$ ◯ $\frac{3}{6}$

3.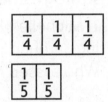

| $\frac{1}{4}$ | $\frac{1}{4}$ | $\frac{1}{4}$ |

| $\frac{1}{5}$ | $\frac{1}{5}$ |

$\frac{3}{4}$ ◯ $\frac{2}{5}$

4. $\frac{3}{8}$ ◯ $\frac{1}{4}$

5. $\frac{2}{3}$ ◯ $\frac{5}{6}$

6. $\frac{4}{8}$ ◯ $\frac{3}{6}$

Problem Solving and TAKS Prep

USE DATA For 7–8, use the model.

7. Whose house is closer to school, Todd's or Al's?

_____

8. Dan walked from his house to school and then to Todd's house. Which distance is farther?

_____

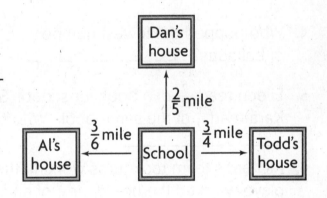

9. I am greater than $\frac{2}{8}$ and less than $\frac{5}{6}$. My denominator is 2. What fraction am I?

A $\frac{1}{2}$

B $\frac{0}{2}$

C $\frac{3}{8}$

D $\frac{2}{2}$

10. Which fraction is greater than $\frac{3}{5}$?

F $\frac{3}{6}$

G $\frac{1}{4}$

H $\frac{7}{8}$

J $\frac{6}{10}$

Practice

Problem Solving Workshop Strategy:
Compare Strategies

Choose a strategy. Then solve.

1. Lisa and Michelle played a ring toss game at the carnival. Lisa tossed $\frac{2}{3}$ of 18 rings around the bottle. Michelle tossed $\frac{5}{6}$ of 18 rings around the bottle. Who tossed more rings around the bottle? _____

2. Chris and his friends ordered ice cream sundaes at the food stand. Chris ate $\frac{1}{3}$ of his sundae. Hayden ate $\frac{3}{5}$ of his sundae and Jacob ate $\frac{2}{8}$ of his sundae. Who ate the most of his sundae? _____

Mixed Strategy Practice

USE DATA For 3–4, use the table.

3. For the Balloon Pop game, players have chances to pop 8 balloons. Who popped the greatest number of balloons? _____

Balloon Pop Game	
Name of Player	**Fraction of Balloons Popped**
Taylor	$\frac{5}{8}$
Sean	$\frac{3}{4}$
Roseanne	$\frac{1}{2}$

4. Who popped the fewest number of balloons? _____

5. Eileen read $\frac{7}{10}$ of a book for school. Shelly read $\frac{4}{5}$ of the same book and Kara read $\frac{1}{2}$ of the same book. Who read the most of the book? _____

6. Richard's team took turns working the scoreboard at the game. Every player worked the board for $\frac{1}{6}$ of an hour. The game lasted for 2 hours. How many players are on the team? _____

Practice

Arrays with Tens and Ones

Find the product.

1.

$2 \times 16 =$ _____

2.

$4 \times 13 =$ _____

3.

$3 \times 22 =$ _____

4.

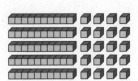

$5 \times 14 =$ _____

5.

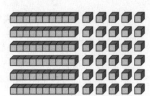

$6 \times 15 =$ _____

6.

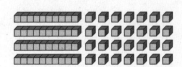

$4 \times 17 =$ _____

Use base-ten blocks or grid paper to find each product.

7. $5 \times 25 =$ _____

8. $4 \times 18 =$ _____

9. $4 \times 22 =$ _____

10. $3 \times 19 =$ _____

11. $4 \times 27 =$ _____

12. $8 \times 39 =$ _____

13. $6 \times 88 =$ _____

14. $4 \times 12 =$ _____

15. $7 \times 31 =$ _____

16. $3 \times 24 =$ _____

17. $4 \times 29 =$ _____

18. $9 \times 15 =$ _____

19. $8 \times 16 =$ _____

20. $5 \times 35 =$ _____

© Harcourt

Model 2-Digit Multiplication

Find the product. Use place value or regrouping.

1.
25
× 2

2.
16
× 4

3.
34
× 3

Multiply. You may wish to use base-ten blocks to help you.

4. 22
    × 7

5. 36
    × 3

6. 43
    × 5

7. 24
    × 6

8. 32
    × 5

9. 18
    × 4

10. 31
     × 4

11. 16
     × 4

Use base-ten blocks to find the missing factor.

12. $3 \times \square = 96$
13. $\square \times 4 = 56$
14. $\square \times 38 = 76$
15. $5 \times \square = 90$

Problem Solving and TAKS Prep

16. There are 300 brushes in a pack. Ella bought 4 packs. How many brushes did Ella buy?

_____

17. There are 20 boxes of crayons in a case. If each box costs $3 how much does a case cost?

_____

18. Carter's class went on a picnic. There were 13 students in each of 4 groups. How many students went on the picnic?

   A 48
   B 52
   C 56
   D 60

19. Eddie reads 2 hours a day. How many hours does he read in 12 weeks?

   F 14
   G 7
   H 168
   J 100

Practice

● Multiply 2-Digit Numbers

Find each product.

1. 23
   × 4

2. 78
   × 6

3. 77
   × 6

4. 15
   × 9

5. 34
   × 7

6. 39
   × 7

7. 92
   × 3

8. 41
   × 7

9. 84
   × 2

10. 67
    × 3

11. $95 \times 8 =$ _____

12. $57 \times 6 =$ _____

13. $4 \times 99 =$ _____

14. $6 \times 73 =$ _____

Problem Solving and TAKS Prep

USE DATA For 15–16, use the graph.

15. What number times 3, minus 16, equals the number of types of lunches sold in all?

_____

16. What number times 10 equals the number of types of lunches sold in all

_____

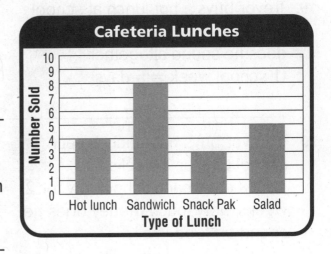

Cafeteria Lunches

Number Sold / Type of Lunch
Hot lunch   Sandwich   Snack Pak   Salad

17. Vincent uses 48 inches of wood to make a frame. How many inches of wood will Vincent need, to make 9 frames?

   A 475 inches

   B 540 inches

   C 432 inches

   D 480 inches

18. Colleen listened to three CDs. Each CD is 63 minutes long. How many minutes did it take for Colleen to listen to all three CDs?

   F 146 minutes

   G 169 minutes

   H 189 minutes

   J 378 minutes

Practice

Practice 2-Digit Multiplication

Multiply. Use partial products or regrouping.

1.	2.	3.	4.	5.
23 × 7	78 × 3	28 × 2	53 × 4	34 × 7

6.	7.	8.	9.	10.
33 × 2	67 × 5	52 × 9	82 × 3	71 × 5

11. $95 \times 4 =$ _____ **12.** $57 \times 5 =$ _____ **13.** $4 \times 39 =$ _____ **14.** $2 \times 77 =$ _____

15. $32 \times 5 =$ _____ **16.** $9 \times 15 =$ _____ **17.** $3 \times 21 =$ _____ **18.** $57 \times 8 =$ _____

Problem Solving and TAKS Prep

USE DATA For 19–20, use the Menu.

19. Trevor buys a hot lunch at school every day for 2 weeks. How much does he spend altogether?
(1 school week = 5 days)

_____

20. If Trevor buys hot lunches each school day for 2 weeks and sandwiches each school day for 2 weeks, how much money does he spend on food in all?

_____

School Lunch Menu

Hot lunch		$2
Sandwich		$3
Snack Pak		$5
Salad		$4

21. Andrew collected 32 planks of wood to make a frame. If he uses the same number of planks to make each frame, how many planks of wood will he need to make 8 frames?

A 266 planks
B 256 planks
C 326 planks
D 156 planks

22. Jackie listened to 4 CDs. Each CD was 42 minutes long. How long did it take Jackie to listen to all three CDs?

F 146 minutes
G 169 minutes
H 168 minutes
J 378 minutes

Practice

© Harcourt

Problem Solving Workshop Strategy:
Solve a Simpler Problem

Problem Solving Strategy Practice

1. The music club gives a concert to raise money for new sheet music. They get $0.75 for each ticket they sell. The club sold 99 tickets. How much money did they raise?

   _____

2. Brett uses toy houses to build his model village. The toy houses come in packages of 65. If Brett buys 4 packages, how many houses will he have?

   _____

Mixed Strategy Practice

USE DATA For 3–4, use the chart.

3. Third grade classes collected canned goods for a food drive. The table shows the foods they collected. If 4 third grade classes each collected the same number of cans of each type of food, what is the total number of cans of peas and corn they collected?

   _____

Food	Number of cans
Peas	35
Corn	27
Chicken	15
Potatoes	22
Soup	13

4. If the 4 third grade classes each collected the same number of each type of food, how many more cans of potatoes did they collect than chicken?

   _____

5. Heath volunteers at a library 3 days a week every summer. He reshelves 97 or more books each day he volunteers. What is the least number of books he reshelves each week?

   _____

6. A brown, a black, a white, and a gray dog are in line at a training class. The black dog is not last. The white dog is in front of the brown dog. The brown dog is second. Draw a picture to show the order of the dogs.

Probability: Likelihood of Events

For 1–6, use the bag of tiles. Each tile is the same size and shape.
Tell whether each event is *more likely, less likely, certain*, or *impossible*.

1. pulling a blue tile

_____

2. pulling a red tile

_____

3. pulling a white tile

_____

4. pulling a yellow tile

_____

5. pulling a tile

_____

6. pulling a green, blue, yellow, or red tile

_____

B is blue   **G** is green
R is red   **Y** is yellow

Problem Solving and TAKS Prep

USE DATA For 7–8, use the table. Ben pulls one prize from the bag
without looking. Each prize is the same size and shape.

7. Is it certain or impossible that Ben will pull a stuffed toy?

_____

8. Is it more likely or less likely that Ben will pull a red ball?

_____

Prize Bag	
Prizes	**Number**
blue ball	3
red ball	5
green ball	1

9. Charles grabs a shirt to wear from his drawer without looking. Four of his shirts are white, 1 is yellow, and 5 are blue. Which represents the likelihood that Charles grabs a yellow shirt, if all of the shirts are the same size?

  A more likely   **C** certain

  B less likely   **D** impossible

10. Sara is playing a game using a spinner. The spinner contains 8 sections of equal size: 1 green, 3 blue, 2 white, and 2 red. Which color is the spinner least likely to land on?

  F green   **H** white

  G blue   **J** red

© Harcourt

Practice

Possible Outcomes

For 1–2, list the possible outcomes for each.

1. Elizabeth will pull a marble from the bag.

_____

_____

_____

R is red   G is green
B is blue

2. Joe will use the spinner.

_____

_____

_____

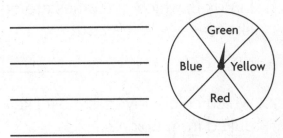

Problem Solving and TAKS Prep

USE DATA For 3–4, use the spinner with equal sized sections below.

3. John is going to use the spinner. What are the possible outcomes?

_____

_____

4. If John spins the spinner one time, is it equally likely that it will land on green or orange?

_____

5. Which is NOT a possible outcome of spinning a spinner with the following colors one time: yellow, green, blue?

   A yellow

   B white

   C blue

   D green

6. Which are equally likely outcomes for a spinner, with sections of equal size, with these sections: 2 yellow sections, 3 red sections, 4 white sections, and 2 blue sections?

   F yellow and red

   G yellow and white

   H yellow and blue

   J red and white

Practice

Experiments

For 1–3, use the boxes of crayons. Each crayon is the same size and shape.

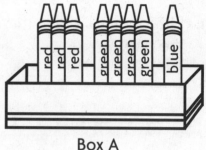

Box A

1. In Box B, which outcomes are equally likely?

   _____

2. Which color crayon is most likely to be pulled from Box A?

   _____

3. What are the possible outcomes for Box A?

   _____

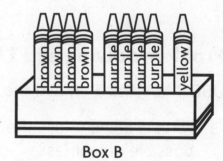

Box B

Problem Solving and TAKS Prep

4. A box of cookies, each of equal size and shape, contains 4 raisin, 4 oatmeal, and 6 ginger cookies. Which cookie is most likely to be pulled in one pull?

   _____

5. Which outcomes are equally likely for a bag of marbles, each of equal size, with 2 red, 3 green, and 2 yellow marbles?

   _____

6. Which outcome is least likely for a bag of marbles, each of equal size, with 4 red, 2 blue, 1 green, and 3 yellow marbles?

   A red   **C** green

   B blue   **D** yellow

7. Which is the probability of pulling a green marble from a bag, with marbles of equal size, of 4 red, 2 blue, 1 green, and 3 yellow marbles?

   F 1 out of 10   **H** 3 out of 10

   G 2 out of 10   **J** 4 out of 10

Practice

Predict Future Events

1. The tally table below shows the results of 30 pulls from a bag of marbles, of equal size. Use the data in the tally table to predict the color that will be pulled next. Explain.

Marble Results	
Color	Tally
Yellow	ЖЖ ЖЖ IIII
Orange	ЖЖ ЖЖ
Purple	ЖЖ I

_____

_____

2. The tally table below shows the results of 25 spins of a spinner, with equal sized sections. Use the data in the tally table to predict the color the pointer will land on next. Explain.

Spinner Results	
Color	Tally
Red	ЖЖ II
Green	ЖЖ I
Blue	ЖЖ ЖЖ II

_____

_____

Problem Solving and TAKS Prep

3. **USE DATA** The graph shows the results of 35 pulls from a bag of tiles. William is going to pull a tile from the bag. Predict the color he will pull.

_____

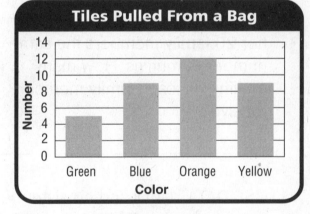

4. The tally table gives the results of 25 pulls from a bag of marbles. Which outcome occurred most often?

A red
B blue
C yellow
D green

Marble Results	
Color	Tally
Red	ЖЖ
Blue	ЖЖ ЖЖ I
Yellow	ЖЖ IIII

5. The tally table shows the results of tossing a coin 30 times. Which is the best prediction for the next coin toss?

F Heads
G Tails
H Neither
J Both are equally likely

Side	Tally
Heads	ЖЖ ЖЖ ЖЖ
Tails	ЖЖ ЖЖ ЖЖ

Problem Solving Workshop Strategy:
Make an Organized List

Problem Solving Strategy Practice

USE DATA For 1–2, use the table.

1. Peter wants to make a sandwich with 1 type of meat and 1 type of bread. How many different 1 meat and 1 bread combinations can Peter make?

Meat	Cheese	Bread
roast beef	swiss	white
turkey	cheddar	wheat
ham		

_____

2. Lizzy wants to make a sandwich with 1 type of bread and 1 type of cheese. How many different 1 bread and 1 cheese combinations can Lizzy make?

_____

Mixed Strategy Practice

3. Mara baked 60 muffins for a family gathering. She wants to give each of her 20 family members the same number of muffins. How many muffins will each family member receive?

4. Frank has 15 pennies, 6 dimes, and 5 nickels. How many different combinations of change can Frank make, so that he can buy a piece of gum that costs $0.23?

_____

5. **USE DATA** Jamal and his sister need school supplies. They each need 8 pencils, 10 markers, 10 pens, and 2 folders. How many packs of each type of school supply do Jamal and his sister need to buy? Draw a diagram to help solve.

School Supplies	
Type	Number per Pack
pencils	12
markers	8
pens	10
folders	4

Practice

SPIRAL
REVIEW

Spiral Review

For 1–5, write the value of
the underlined digit.

1. 9,4<u>2</u>0   _____

2. <u>1</u>,609   _____

3. 2,09<u>3</u>   _____

4. <u>3</u>,826   _____

5. 7,<u>8</u>24   _____

For 6–7, read the thermometer.
Write the temperature.

6.

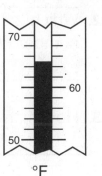
_____ °F

°F

7.

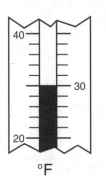
_____ °F

°F

For 8–9, a class takes a
survey about pets. Write the
results as tally marks.

8. 3 students have dogs.

_____

9. 6 students have fish.

_____

10. Look at
the table
at the
right.
How
many
students
were
absent on
Monday?

_____

Absences	
Day	**Tally**
Monday	✝✝✝✝
Tuesday	I
Wednesday	III

For 11–14, draw the next
shape in the pattern.

11. ■ □ ■ □   _____

12. ▯ ○ ○ ▯ ○ ○   _____

13. ⬭ ❘ ❘ ⬭ ❘ ⬭ ❘ ❘   _____

14. ◺ ◹ ◺ ◹ ◹   _____

© Harcourt

Spiral Review

For 1–5, write the value of the underlined digit.

1. 7,8<u>1</u>6 _____

2. <u>9</u>,217 _____

3. 6,42<u>2</u> _____

4. <u>3</u>,405 _____

5. 6,<u>2</u>12 _____

For 6–10, write the time shown on the clock.

6.  _____

7. _____

8. _____

9.  _____

10.  _____

For 11–13, use the graph to answer the questions.

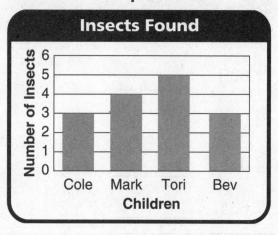

11. Who found the most insects?

_____

12. Which 2 children found the same number of insects?

_____

13. How many insects did mark find?

_____

For 14–17, draw a line on the figure to make the given shapes.

14. Make a triangle and a trapezoid.

15. Make two triangles.

16. Make two rectangles.

17. Make a triangle and a trapezoid.

Spiral Review

Spiral Review

For 1–5, compare. Use <, >, or = for each ◯.

1. 546 ◯ 748

2. 208 ◯ 200

3. 969 ◯ 996

4. 6,399 ◯ 6,399

5. 3,000 ◯ 2,999

For 6–8, give the area of the figure.

6.   _____ square units

7. 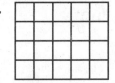 _____ square units

8. 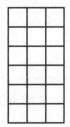 _____ square units

For 9–11, Use the tally table below. Let the 🧍 symbol equal 1 student. Draw the number of symbols needed to show the data for each sport.

Favorite Sport	
Sport	Tally
Swimming	IIII
Karate	IIIII I
Soccer	I

9. Swimming   _____

10. Karate   _____

11. Soccer   _____

For 12–15, write the missing numbers in the pattern.

12. Skip count by twos:

16, 18, ____,____,____,____

13. Skip count by threes:

31, 34, ____,____,____,____

14. Skip count by fives:

55, 60, ____,____,____,____

15. Skip count by tens:

49, 59, ____,____,____,____

Spiral Review

For 1–5, find each difference. Use addition to check.

1.	536 −159	**2.**	627 −548
3.	972 −601	**4.**	840 −588
5.	900 −199		

For 10–12, use the pattern in the table to answer the questions.

number of spiders	1	2	3	4		6
number of legs	8	16	24		40	

10. How many legs do 4 spiders have?

_____

11. How many spiders have 40 legs?

_____

12. How many legs do 6 spiders have?

_____

For 6–9, shade the given area.

6. 6 square units

7. 10 square units

8. 9 square units

9. 13 square units

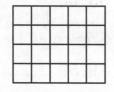

For 13–15, use the number lines to find the missing numbers.

13.

0 1 2 3 4 5 6 7 8 9 ?

_____

14.
0 1 2 3 4 5 6 ? 8 9 10

_____

15.
0 1 ? 3 4 5 6 7 8 9 10

_____

© Harcourt

Spiral Review

Spiral Review

For 1–5, round each number to the nearest hundred.

1. 1,580 _____

2. 2,094 _____

3. 6,527 _____

4. 9,099 _____

5. 602 _____

For 9–11, a class takes a survey about how they come to school. Write the results as tally marks.

9. 7 students ride bikes to school.

_____

10. 11 students ride the bus.

_____

11. Look at the table at the right. How many students went on vacation over Spring Break?

_____

Spring Break	
Activity	**Tally**
Stay Home	⊬⊬⊬ |||
Visit Family	⊬⊬⊬	
Vacation	⊬⊬⊬ ⊬⊬⊬

For 6–8, fill in the blank. A small paper clip is about 1 inch long.

6. _____

This line is about _____ inches long.

7. _____

This line is about _____ inches long.

8. _____

This line is about _____ inch long.

For 12–15, write the next number in the pattern.

12. 1, 3, 5, 7, 9, _____

13. 27, 22, 17, 12, 7, _____

14. 5, 8, 11, 14, 17, _____

15. 10, 20, 30, 40, 50, 60, _____

Spiral Review

For 1–5, use the models to find the difference.

1. 400
    −263

2. 210
    −193

3. 316
    −144
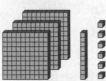

4. 142
    −99
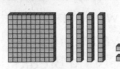

5. 405
    −158
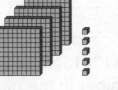

For 6–8, fill in the blank. A child's finger is about 1 cm wide.

6. _____

    This line is about _____ cm long.

7. _____

    This line is about _____ cm long.

8. _____

    This line is about _____ cm long.

For 9–11, use the graph to answer the questions.

9. What beverage was served most often?

    _____

10. How many apple juices were served? _____

11. How many more lemonades were served than milk? _____

For 12–13, use the number line to find the missing numbers.

12. ← | 0 1 2 3 4 5 6 7 8 9 10 11 ? →

    _____

13. ← | 0 1 2 3 ? 5 6 7 8 9 10 11 12 →

    _____

Spiral Review

Spiral Review

For 1–5, use rounding or compatible numbers to estimate each sum.

1. 47 + 52

2. 28 + 23

3. 576 + 139

4. 304 + 188

5. 146 + 149

For 6–7, read the thermometer. Write the temperature.

6.

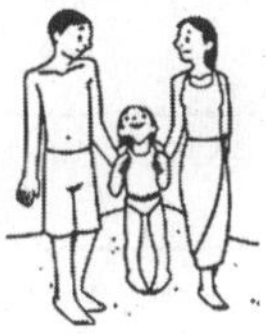

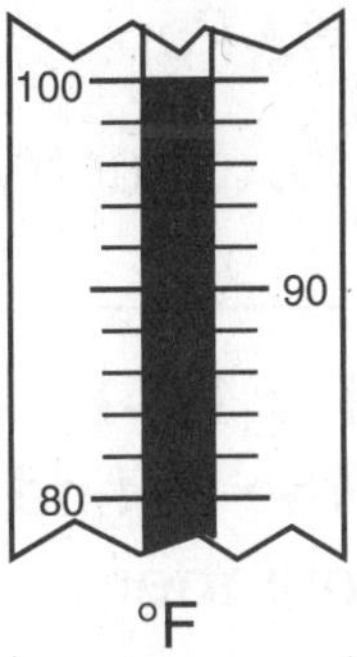

______°F

7.

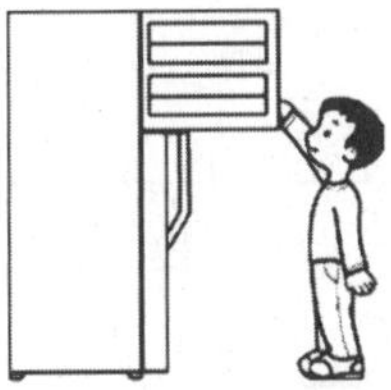

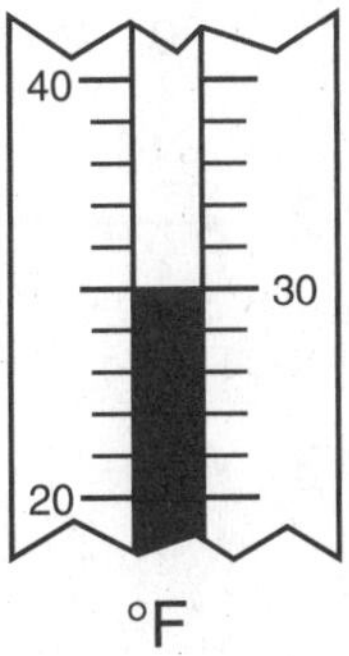

______°F

For 8–9, a class takes a survey about recess. Write the results as tally marks.

8. 8 students play basketball. ________

9. 4 students catch bugs. ________

10. Look at the table below. How many students brought sack lunches on Wednesday?

____

Sack Lunches Brought

Day	Tally
Monday	卌 卌
Tuesday	卌 |
Wednesday	卌 |||

For 11–13, predict the next number in each pattern. Explain.

11. 1, 2, 4, 8, ☐

____

12. 19, 23, 27, 31, ☐

____

13. 900, 800, 700, 600, ☐

____

Spiral Review

For 1–3, write the amount.

 1.

_____

2.

_____

3.

_____

For 4–7, shade the given area.

4. 8 square units

5. 15 square units

6. 4 square units

7. 18 square units

For 8–10, use the tally table below. Let the 𝑋 symbol equal 1 student. Insert number of symbols needed to show the data.

What Music We Listen To	
Music	**Tally**
Rap	
Pop	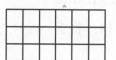
Country	

8. Rap _____

9. Pop _____

10. Country _____

For 11–13, use the number line to find the missing numbers.

11.
0 2 4 6 ? 10 12 14 16 18 20

12.
0 2 4 6 8 10 12 14 ? 18 20

13.
0 2 ? 6 8 10 12 14 16 18 20

Spiral Review

Spiral Review

For 1–5, compare.
Use <, >, or = for each ◯.

1. 1,558 ◯ 1,558

2. 7,094 ◯ 7,904

3. 848 ◯ 8846

4. 3,547 ◯ 3,547

5. 4,999 ◯ 5,001

For 6–10, write the time shown on the clock.

6. _____

7.  _____

8.  _____

9.  _____

10.  _____

For 11–12, a class takes a survey about favorite colors. Write the results as tally marks.

11. 13 students chose red.

12. 9 students chose blue.

_____   _____

13. Look at the table below. How many more students wore white shirts than green shirts?

_____

What Color Shirt?	
Color	**Tally**
White	ⅢⅢ Ⅰ
Orange	‖
Green	‖‖

For 14–16, predict the next number in each pattern. Explain.

14. 22, 26, 30, 34, 38, ☐

_____

15. 99, 88, 77, 66, ☐

_____

16. 300, 350, 400, 450, ☐

_____

Spiral Review

For 1–5, find each sum. Use subtraction to check.

1. 567
   +207

2. 789
   +116

3. 207
   +718

4. 836
   +855

5. 207
   +793

For 6–10, write the time shown on the clock.

6.   _____

7.  _____

8.   _____

9.  _____

10.  _____

For 11–13, use the graph to answer the questions.

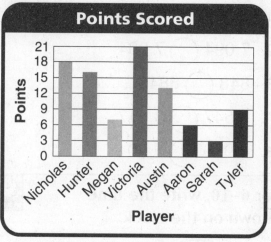

Points Scored

11. Which player scored the most points?

   _____

12. How many points did Nicholas score?

   _____

13. How many more points did Tyler score than Sarah?

   _____

For 14–15, use the number line to find the missing numbers.

14.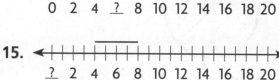

   0 2 4 ? 8 10 12 14 16 18 20

15. _____

   ? 2 4 6 8 10 12 14 16 18 20

   _____

© Harcourt

Spiral Review

For 1–5, use the models to find the difference.

1. 290
   −217

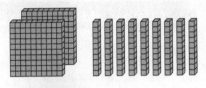

2. 401
   −158

3. 250
   −167

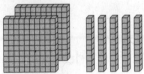

4. 202
   − 79

5. 320
   −131

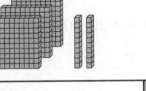

For 9–11, use the tally table below. Let the 👤 symbol equal one student. Draw the number of symbols needed to show the data for each subject.

Favorite Subject	
Subject	Tally
Math	ЖЖ Ⅰ
Science	ЖЖ ⅠⅠ
Reading	ЖЖ

9. Math _____

10. Science _____

11. Reading _____

For 6–8, draw the time on the clock.

6.

 9:41

7.

 10:38

8.

 2:23

For 12–14, draw the next shape in the pattern.

12. □ ○ □ □ ○ □ ○ ___

13. ‖ • ‖ • ‖ • Ⅰ ___

14. □   □□   □□ ___

Spiral Review

For 1–5, round each number to the nearest hundred.

1. 6,581 _____

2. 1,157 _____

3. 8,502 _____

4. 8,205 _____

5. 495 _____

For 6–9, shade the given area.

6. 12 square units

7. 14 square units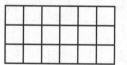

8. 7 square units

9. 15 square units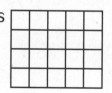

For 10–12, use the graph to answer the questions.

Number of Pushups in P.E. Class

10. Which student did exactly 16 pushups? _____

11. How many more pushups did Connor do than Michael? _____

12. How many students did more than 20 pushups? _____

For 13–16, draw a line on the figure to make the given shapes.

13.   Make two trapezoids.

14.   Make two triangles.

15.   Make two squares.

16.   Make two triangles.

Spiral Review

For 1–5, use the pictures to find the missing number.

1. ☐ × 5 = 10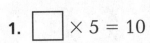

2. ☐ × 3 = 9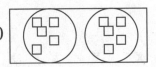

3. ☐ × 2 = 8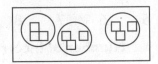

4. ☐ × 9 = 9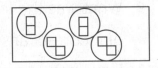

For 5–7, fill in the blank. A small paper clip is about 1 in. long.

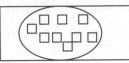

5.

This string is about ____ inch long.

6.)Crayon(

This crayon is about ____ inches long.

7.

This leaf is about ____ inches long.

For 8–9, a class takes a survey about their heroes. Write the results as tally marks.

8. 15 students chose Martin Luther King Jr.

_____

9. 9 students chose Eleanor Roosevelt.

_____

10. Look at the table at the right. How many more students chose Mother Teresa than Thomas Edison? _____

Who is Your Hero?	
Person	**Tally**
Thomas Edison	卌 ‖
Abraham Lincoln	‖‖
Mother Teresa	卌 ‖‖

For 11–14, find the product.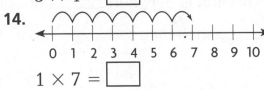

11.

```
←+--+--+--+--+--+--+--+--+--+--→
  0  1  2  3  4  5  6  7  8  9  10
```

2 × 5 = ☐

12.

```
←+--+--+--+--+--+--+--+--+--+--→
  0  1  2  3  4  5  6  7  8  9  10
```

3 × 2 = ☐

13.

```
←+--+--+--+--+--+--+--+--+--+--→
  0  2  4  6  8  10 12 14 16 18 20
```

5 × 4 = ☐

14.

```
←+--+--+--+--+--+--+--+--+--+--→
  0  1  2  3  4  5  6  7  8  9  10
```

1 × 7 = ☐

Spiral Review

For 1–5, use rounding or compatible numbers to estimate each sum.

1. $\begin{array}{r} 36 \\ +19 \\ \hline \end{array}$	**2.** $\begin{array}{r} 61 \\ +85 \\ \hline \end{array}$
3. $\begin{array}{r} 306 \\ +294 \\ \hline \end{array}$	**4.** $\begin{array}{r} 919 \\ +237 \\ \hline \end{array}$
5. $\begin{array}{r} 278 \\ +144 \\ \hline \end{array}$	

For 9–11, use the graph to answer the questions.

```
                    X     X
   X                X  X  X
   X  X  X  X  X  X  X
   +--+--+--+--+--+--+--+
   5  6  7  8  9 10  11
```
Number of Incorrect Answers

9. What was the least number of incorrect answers? _____

10. How many people answered exactly 7 questions incorrectly?

_____

11. What was the greatest number of incorrect answers? _____

For 6–8, fill in the blank. A small paper clip is about 1 in. long.

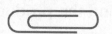

6. _____

This line is between _____ and _____ inches long.

7. _____

This line is between _____ and _____ inches long.

8. _____

This line is between _____ and _____ inch long.

For 12–15, draw a line on the figure to make the given shapes.

12. Make two triangles.

13. Make two rectangles.

14. Make two trapezoids.

15. Make one triangle and one trapezoid.

Spiral Review

Spiral Review

For 1–3, write the amount.

1.  _____

2.  _____

3.  _____

For 8–9, a class takes a survey about their favorite type of movie. Write the results as tally marks.

8. 13 students like comedy.

9. 7 students like drama.

_____   _____

10. Look at the table below. How many more students prefer puzzle than action video games?

_____

Favorite Video Game	
Type	Tally
Action	ⅡⅡⅡ Ⅰ
Fantasy	ⅡⅡⅡ
Puzzle	ⅡⅡⅡ ⅠⅠⅠ

For 4–7, name the shaded area.

4. _____ square units

5. 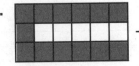 _____ square units

6. _____ square units

7.  _____ square units

For 11–14, find the product.

11. $3 \times 4 = \boxed{}$

12. $3 \times 6 = \boxed{}$

13. $4 \times 5 = \boxed{}$

14. $8 \times 2 = \boxed{}$

Spiral Review

For 1–5, use the pictures to find the missing factor.

1. ☐ × 4 = 8

2. ☐ × 4 = 12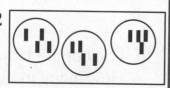

3. ☐ × 1 = 4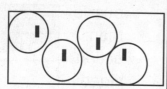

4. ☐ × 12 = 12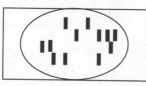

For 7–9, use the graph to answer the questions.

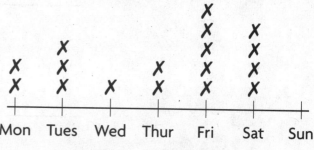

Number of Smoothies Sold

7. How many more smoothies were sold on Friday than on Monday?____

8. On what day were exactly 4 smoothies sold? _____

9. On what day was the store probably closed? _____

For 5–6, read the thermometer. Write the temperature.

5.

_____ °F

6.

_____ °F

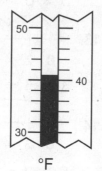

For 10–12, use the number line to find the missing numbers.

10.
0 1 2 3 4 5 6 7 ? 9 10 11 12 13 14 15 16 17 18 19 20

_____

11.
0 1 2 3 4 5 6 7 8 9 10 11 ? 13 14 15 16 17 18 19 20

_____

12.
0 1 2 3 4 5 6 7 8 9 10 11 12 13 14 15 16 ? 18 19 20

_____

Spiral Review

Spiral Review

For 1–5, write the numbers in order from least to greatest.

1. 707, 139, 610, 601

   _____

2. 475, 919, 199, 105

   _____

3. 1,978; 2,559; 1,879; 1,421

   _____

4. 2,228; 3,366; 3,334; 2,316

   _____

5. 7,845; 7,942; 7,930; 7,854

   _____

For 9–13, change the numbers to tally marks.

Rose Varieties Sold		
Variety	Number	Tally
9. Belinda's Dream	7	
10. Abraham Darby	4	
11. Darland	11	
12. Knockout	19	
13. Joseph's Coat	5	

For 6–8, draw the time on the clock.

6.

7.

8.

For 14–17, use the arrays to find the product.

14. $7 \times 2 = \boxed{}$

15. $8 \times 4 = \boxed{}$

16. $9 \times 3 = \boxed{}$

17. $5 \times 6 = \boxed{}$

Spiral Review

Spiral Review

**For 1–5, find each sum.
Use subtraction to check.**

1. 916
   +450

2. 507
   +589

3. 954
   +647

4. 784
   +169

5. 109
   +317

**For 10–12, use the graph
to answer the questions.**

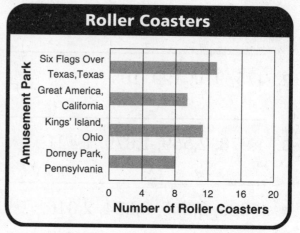

Roller Coasters

Amusement Park:
Six Flags Over Texas, Texas
Great America, California
Kings' Island, Ohio
Dorney Park, Pennsylvania

Number of Roller Coasters
0 4 8 12 16 20

10. How many more roller coasters
    does King's Island have
    than Great America? _____

11. Which park has exactly 8
    coasters?

    _____

12. Which park has more than 12
    coasters?

    _____

**For 6–9, shade the
given area.**

6. 13 square
   units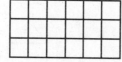

7. 15 square
   units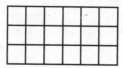

8. 8 square
   units

9. 16 square
   units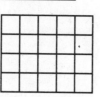

**For 13–16, write the
number of vertices.**

Figure	Vertices
13.	____
14.	____
15.	____
16.	____

Spiral Review

For 1–5, compare.
Write <, >, or = for each ◯.

1. 7,615 ◯ 7,651

2. 4,507 ◯ 4,507

3. 749 ◯ 794

4. 3,518 ◯ 3,509

5. 2,450 ◯ 2,405

For 6–8, fill in the blank.
A child's finger is about 1 centimeter wide.

6.

This string is between _____ and

_____ centimeters long.

7.

This crayon is between _____

and _____ centimeters long.

8.

This leaf is between _____ and

_____ centimeters long.

For 9–13, a class takes a survey about their favorite juices. Write the results as tally marks.

Favorite Juice		
Flavor	Number	Tally
9. Grape	16	
10. Orange	6	
11. Strawberry-Banana	5	
12. Apple	3	
13. Cranberry	10	

For 14–17, use the array to find the product.

14. $4 \times 6 =$ ▢

15. $7 \times 5 =$ ▢

16. $2 \times 6 =$ ▢

17. $9 \times 4 =$ ▢

Name _____

Spiral Review

For 1–3, write a division sentence for each model.

1.

   _____

2.

   _____

3.

   _____

For 4–6, fill in the blank.
A child's finger is about 1 centimeter wide.

4. _____

   This line is between _____ and _____ centimeters long.

5. _____

   This line is between _____ and _____ centimeters long.

6. _____

   This line is between _____ and _____ centimeters long.

For 7–9, use the graph to answer the questions.

Amusement Park Rides

7. How many parks have over 60 rides? _____

8. Which park has exactly 60 rides?
   _____

9. Which parks have the same number of rides?
   _____

For 10–13, write the number of vertices.

Figure	Vertices
10.	_____
11.	_____
12.	_____
13.	_____

Spiral Review

© Harcourt

Spiral Review

For 1–3, write a division
sentence for each model.

1. _____

2. 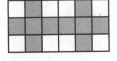 _____

3. _____

For 4–7, name the shaded
area.

4.

_____ square units

5.

_____ square units

6.

_____ square units

7.

_____ square units

For 8–12, read the steps for
making a field trip pictograph. Then
write them in the correct order.

A. Show the correct   **8.** _____
    number of pictures
    beside each field
    trip choice.

B. Write a label for   **9.** _____
    each row.

C. Choose a key to   **10.** _____
    tell how many
    each picture
    stands for.

D. Choose a title.   **11.** _____

E. Decide how   **12.** _____
    many pictures
    should be placed
    next to each field
    trip choice.

For 13–17, write the fact
family for each set
of numbers.

13. 7, 4, 28 _____ _____
                _____ _____

14. 5, 7, 35 _____ _____
                _____ _____

15. 8, 2, 16 _____ _____
                _____ _____

16. 9, 3, 27 _____ _____
                _____ _____

17. 6, 6, 36 _____ _____

© Harcourt

Spiral Review

For 1–5, find the product.

1. 26
  × 8

2. 54
  × 6

3. 71
  × 9

4. 68
  × 3

5. 87
  × 2

For 8–10, use the graph to answer the questions.

Number of National Parks						
Massachusetts	♀	♀	♀	♀	♀	
Michigan	♀	(				
New Jersey	♀	♀	(			
New York	♀	♀	♀	♀	♀	♀ ♀
Pennsylvania	♀	♀	♀	♀	♀	♀
Key: Each ♀ = 4 national parks						

8. Which state has the least number of National Parks? _____

9. How many more National Parks does Pennsylvania have than Massachusetts? _____

10. Which state has 28 National Parks? _____

For 6–7, read the thermometer. Write the temperature.

6.

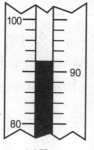

_____ °F

7.

_____ °F

For 11–13, draw a line to match the figure with the shape of its face or faces.

11. □

12. □

13. ▭

© Harcourt

Spiral Review

For 1–5, find the product.

1. 19
   ×7

2. 34
   ×8

3. 65
   ×4

4. 54
   ×6

5. 28
   ×5

For 9–11, use the tally table.
Let the 👤 equal 2 students. Draw
the number of symbols needed to
show the data for each president.

Favorite President	
President	Tally
Washington	JHT JHT II
Lincoln	JHT I
Kennedy	JHT JHT

9. Washington _____

10. Lincoln _____

11. Kennedy _____

For 6–8, draw the time
on the clock.

6.

7.

8.

For 12–14, use the pattern
in the table to answer the
questions.

number of cars	1	2	3	4		6
number of wheels	4	8	12		20	

12. How many wheels do
    4 cars have? _____

13. How many cars have
    20 wheels? _____

14. How many wheels do
    6 cars have? _____

Spiral Review

For 1–5, use the models to find the difference.

1. 300
    −172

2. 401
    −243

3. 327
    −274

4. 193
    − 88
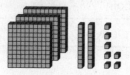

5. 111
    − 49
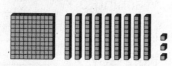

For 6–9, shade the given area.

6. 7 square units

7. 8 square units

8. 9 square units

9. 10 square units

For 10–12, use the graph to answer the questions.

Roller Coaster Speed

Top Thrill Dragster · Nitro · Superman The Escape · Gemini
Name
0   50   100   150
Miles Per Hour

10. About how fast does the Top Thrill Dragster go?

_____

11. Which is the slowest roller coaster?

_____

12. Which roller coaster is faster than Nitro but slower than Top Thrill Dragster?

_____

For 13–16, match the figure to the name.

13.   triangle

14.   trapezoid

15. 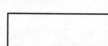   rhombus

16.   rectangle

Spiral Review

Spiral Review

For 1–3, write the amount.

1.

_____

2.

_____

3.

_____

For 4–6, fill in the blank.
A child's finger is about
1 centimeter wide.

4.

This bee is between _____ and

_____ centimeters long.

5.

This caterpillar is between _____

and _____ centimeters long.

6.

This praying mantis is between

_____ and _____ centimeters long.

For 7–8, a class takes a survey about their favorite vacation spots. Write the results as tally marks.

7. 11 students like the beach.

_____

8. 4 students like the mountains.

_____

9. Look at the table below. How many more students prefer amusement parks to mountains? _____

Favorite Vacation Spots	
Type	Tally
Beach	‖‖‖ ‖‖‖	
Mountains	‖‖‖‖
Amusement Park	‖‖‖ ‖‖‖ ‖‖‖ ‖‖

For 10–12, use the table to answer the questions.

The electronics shop had a sale.

Batteries	3	6	9
Money	$5	$10	$15

10. What is the rule for this table?

_____

11. How much would it cost to buy 15 batteries? _____

12. How many batteries can you buy with $30? _____

Spiral Review

For 1–5, round each number to the nearest hundred.

1. 7,467 _____

2. 3,507 _____

3. 9,291 _____

4. 974 _____

5. 3,074 _____

For 6–8, shade the given area.

6. 11 square units

7. 13 square units

8. 12 square units

For 9–11, use the graph to answer the questions.

Favorite Ride

9. What was the least favorite ride?

_____

10. How many people voted for the Ferris wheel? _____

11. How many more people voted for the roller coaster than for the Ferris wheel? _____

For 12–15, draw a shape congruent to the one shown.

12.

13.

14.

15.

Spiral Review

© Harcourt

Spiral Review

For 1–4, use the models to find the missing factors.

1. $1 \times \boxed{} = 10$

2. $3 \times \boxed{} = 12$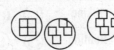

3. $2 \times \boxed{} = 12$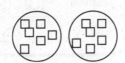

4. $4 \times \boxed{} = 4$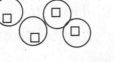

For 5–7, fill in the blank.
A small paper clip is about 1 in. long.

5.

   This pickle is about _____ inches long.

6.

   This green bean is about _____ inches long.

7.

   This pepper is about _____ inch long.

For 8–10, a business takes a survey about their customers. Write the results as tally marks.

8. 8 customers ordered on the Internet.

   _____

9. 6 customers ordered over the phone.

   _____

10. Look at the table at the right. How many customers placed orders in all?

    _____

What Did Your Customers Order?	
Order	Tally
Toy	HHtI
Book	III
CD	HHtIIII

For 11–15, write the related multiplication fact.

11. $9 \times 5 = 45$

    _____

12. $8 \times 4 = 32$

    _____

13. $7 \times 6 = 42$

    _____

14. $9 \times 3 = 27$

    _____

15. $1 \times 10 = 10$

    _____

Spiral Review

For 1–5, use rounding or compatible numbers to estimate each sum.

1. 69
   +34

2. 52
   +29

3. 221
   +670

4. 431
   +369

5. 578
   +24

For 6–7, read the thermometer. Write the temperature.

6.

_____ °F

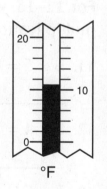

°F

7.

_____ °F

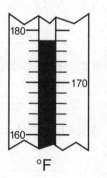

°F

For 8–10, use the graph to answer the questions.

Votes for Favorite Park Activity				
Biking	☺	☺	☺	☾
Hiking	☺	☺	☺	☺
Boating	☺	☺	☺	
Fishing	☺	☾		

Key: Each ☺ = 10 votes.

8. How many more people chose biking than boating? _____

9. For which activity did exactly 15 people vote? _____

10. How many people chose water-related activities? _____

For 11–14, draw a shape with the given line of symmetry.

11.

12.

13.

14.

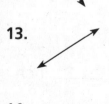

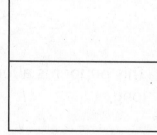

© Harcourt

Spiral Review

For 1–5, find the product.

1. 28
   ×4

2. 19
   ×7

3. 48
   ×3

4. 82
   ×8

5. 53
   ×5

For 6–8, choose the unit you would use to weigh each. Write *ounce* or *pound*.

6.

   _____

7.

   _____

8.

   _____

For 9–11, use the tally table below. Let the 🧍 symbol equal 2 students. Draw the number of symbols needed to show the data.

Schedule Choices	
Choices	**Tally**
5 days a week	ЖЖ ЖЖ
4 days a week	ЖЖ |||
6 days a week	|||

9. 5 days a week _____

10. 4 days a week _____

11. 6 days a week _____

For 12–14, use the pattern in the table to answer the questions.

number of stop signs	1	2	3	4		6
number of sides	8	16	24		40	

12. How many sides do 4 stop signs have? _____

13. How many stop signs have 40 sides? _____

14. How many sides do 6 stop signs have? _____

© Harcourt

Spiral Review

For 1–3, write a division sentence for each model.

1.

2.

3.

For 4–7, match the object with the best unit of measure.

4. distance between two
state capitals   yard

5. length of a sports field   inch

6. width of butterfly wings   mile

7. height of a 1-story building   foot

For 8–10, use the graph to answer the questions.

Number of National Parks Visited	
State	Number
Arizona	5
Colorado	3
Kansas	2
Oregon	2

8. How many National Parks
were visited in Arizona
and Kansas combined? _____

9. Which state has exactly 3
National Parks?

_____

10. How many National Parks
were visited in Kansas
and Oregon combined? _____

For 11–14, draw the line of symmetry on the figure.

11.

12.

13.

14.

© Harcourt

Spiral Review

For 1–5, write the numbers described by the words.

1. Nine thousand nine hundred fifteen _____

2. One thousand seven hundred thirty five _____

3. Five thousand nine hundred fifty seven _____

4. Two thousand eight hundred fifty seven _____

5. Eight thousand nine hundred twenty seven _____

For 6–7, use the thermometers.

7:00 A.M.	3:00 P.M.
60 50 40 °F	60 50 40 °F

6. The temperature rose 6°F from 7:00 A.M. to 8:00 A.M. What was the temperature at 8:00 A.M.?

_____

7. The temperature at 12:00 P.M. was 5°F cooler than the temperature at 3:00 P.M. What was the temperature at 12:00 P.M.?

_____

For 8–12, change the numbers to tally marks.

Phone Calls Received in One Week		
Type	**Number**	**Tally**
8. Wrong Number	2	
9. Sales	6	
10. Call for Mom	12	
11. Call for Dad	9	
12. Call for Kids	16	

For 13–15, use the table to answer the questions.

The bowling alley has a special rate!

Games	2	4	6
Cost	$10	$20	$30

13. What is the rule for this table?

_____

14. How much would it cost to bowl 8 games? _____

15. If you spent $50.00, how many games could you bowl? _____

Spiral Review

Spiral Review

For 1–5, compare the fractions using the number lines.

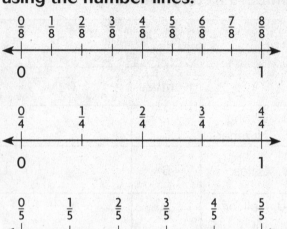

1. $\frac{4}{8} \bigcirc \frac{2}{4}$

2. $\frac{1}{4} \bigcirc \frac{1}{5}$

3. $\frac{3}{5} \bigcirc \frac{5}{8}$

4. $\frac{1}{5} \bigcirc \frac{1}{4}$

5. $\frac{2}{8} \bigcirc \frac{1}{5}$

For 6–9, find the perimeter of each figure.

6. _____ units

7. _____ units

8. _____ units

9. ▢▢▢▢ _____ units

For 10–12, use the graph to answer the questions.

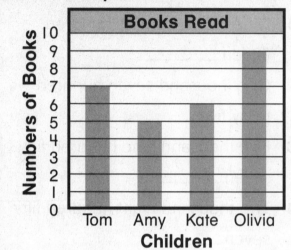

10. Who read two fewer books than Tom? _____

11. Who read fewer books than Kate? _____

12. How many books did the students read in all? _____

For 13–16, match the figure with its name.

13.   pentagon

14. ▭   rectangle

15.   octagon

16.   hexagon

© Harcourt

Spiral Review

For 1–3, shade the amount shown by the fraction.

1. $\frac{3}{4}$

2. $\frac{1}{6}$

3. $\frac{1}{2}$

For 4–6, give the area of the figure.

4.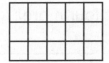

_____ square units

5.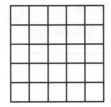

_____ square units

6.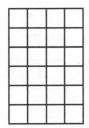

_____ square units

For 7–9, use the tally table below. Let the ☥ symbol equal 3 people. Draw the number of symbols needed to show the data for each book.

Favorite Type of Book	
Type	Tally
Fiction	|||| |||| ||
Non-Fiction	||||	
Biography	|||| ||||

7. Fiction _____

8. Non-Fiction _____

9. Biography _____

For 10–12, use the pattern in the table to answer the questions.

number of songs	1	2	3	4		6
number of verses	5	10	15		25	

10. How many verses are in 4 songs?

_____

11. How many songs have 25 verses?

_____

12. How many verses are in 6 songs?

_____

© Harcourt

Spiral Review

For 1–4, write the fraction for the gray shaded part.

1. _____

2. _____

3.  _____

4. _____

For 5–8, Choose the unit you would use to measure each. Write *cup, pint, quart,* or *gallon*.

5. _____

6. _____

7. _____

8.  _____

For 9–11, use the graph to answer the questions.

Number of Library Books Checked Out				
Gwen	📘	📘	📘	
Tony	📘	📘	📘	📘
Linda	📘	📘		

Key: Each 📘 = 2 books.

9. Who checked out the least number of books? _____

10. How many books were checked out in all? _____

11. Who checked out four more books than Linda? _____

For 12–15, draw a shape congruent to the one shown.

12.

13.

14.

15.

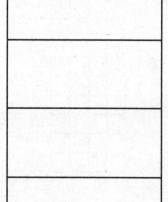

Spiral Review

For 1–3, shade the equivalent fraction. Then write the equivalent fraction.

1.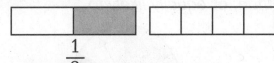

   $\frac{1}{2}$ _____

2.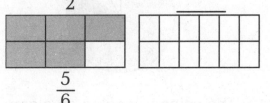

   $\frac{5}{6}$ _____

3.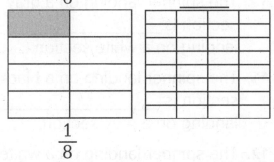

   $\frac{1}{8}$ _____

For 4–6, Find the volume of the solid figures.

4. _____ cubic units

5. _____ cubic units

6. 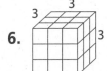 _____ cubic units

For 7–10, complete the sentences with the phrase *less likely than*, *more likely than*, or *equally likely to.*

7. Pulling a black marble is _____
   _____ pulling a gray marble.

8. Pulling a gray marble is _____
   _____ pulling a white marble.

9. Pulling a white marble is _____
   _____ pulling a gray marble.

10. Pulling a penny is _____
    _____ pulling a cotton ball.

For 11–13, use the table to answer the questions.

A landscape company has gardeners planting trees every hour.

Hours	1	2	3
Trees	3	6	9

11. What is the rule for this table?

    _____

12. How long would it take the gardeners to plant 15 trees?

    _____

13. How long would it take the gardeners to plant 24 trees?

    _____

© Harcourt

Spiral Review

For 1–5, compare the fractions using the number lines.

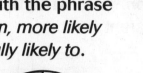

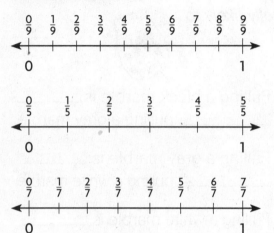

1. $\frac{4}{9} \bigcirc \frac{2}{5}$ 2. $\frac{1}{7} \bigcirc \frac{1}{5}$

3. $\frac{8}{9} \bigcirc \frac{6}{7}$ 4. $\frac{3}{5} \bigcirc \frac{4}{7}$

5. $\frac{2}{9} \bigcirc \frac{3}{5}$

For 6–9, find the perimeter of each figure.

6.   _____ units

7. ⌐⌐   _____ units

8. ▭▭▭   _____ units

9.   _____ units

For 10–13, complete the sentences with the phrase *less likely than, more likely than,* or *equally likely to.*

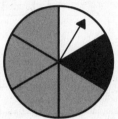

10. The spinner landing on a gray section is _____ landing on a white section.

11. The spinner landing on a black section is _____ landing on a gray section.

12. The spinner landing on a white section is _____ landing on a black section.

13. The spinner landing on a number is _____ landing on a letter.

For 14–17, draw a shape with the given line of symmetry.

14. ↑↓

15. ↖↘

16. ↙↗

17. ⟷